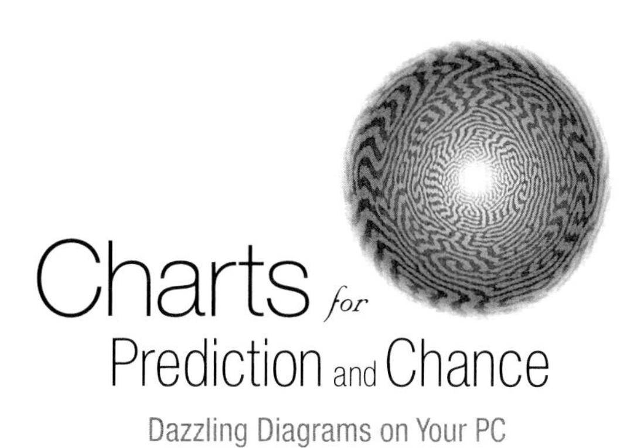
Charts for
Prediction and Chance
Dazzling Diagrams on Your PC

Charts *for* Prediction and Chance

Dazzling Diagrams on Your PC

Mario Markus

Max-Planck-Institut für molekulare Physiologie

Dortmund, Germany

ICP

Imperial College Press

Published by

Imperial College Press
57 Shelton Street
Covent Garden
London WC2H 9HE

Distributed by

World Scientific Publishing Co. Pte. Ltd.

5 Toh Tuck Link, Singapore 596224

USA office: 27 Warren Street, Suite 401-402, Hackensack, NJ 07601

UK office: 57 Shelton Street, Covent Garden, London WC2H 9HE

Library of Congress Cataloging-in-Publication Data
Markus, M. (Mario), 1944-
Charts for prediction and chance : dazzling diagrams on your PC / Mario Markus.
p. cm.
Includes bibliographical references and index.
ISBN-13: 978-1-86094-835-0 (hardcover : alk. paper)
ISBN-10: 1-86094-835-9 (hardcover : alk. paper)
1. Computer science--Mathematics. 2. Computer art. 3. Computer graphics. I. Title.
QA76.9.M35M357 2007
001.4'226028566--dc22

2007027879

British Library Cataloguing-in-Publication Data
A catalogue record for this book is available from the British Library.

Printed in Singapore by Mainland Press Pte Ltd

Preface

This book is directed both to laymen and to scientists, both to technicians and to artists.

Let us start with the scientific and technical domain. Suppose one is dealing with a mathematical, physical, chemical, ecological or economical system, which is controllable by two parameters. These parameters are the coordinates of the diagrams in this book. The grey shadings or colors on them tell us if the system is predictable (like the phases of the moon) or if chance plays a role because its predictability is limited (like the weather).

On the other hand, one may be inclined to draw these diagrams solely for their aesthetic sake. The method is not difficult to understand or program for a user with high school level maths. Moreover using the included CD-ROM removes the need to understand it at all.

I have been drawing diagrams of this kind for a quarter of a century. As a result I have exhibited all over the world, won a prize, and designed the covers of books and CDs as well as logos and postcards. Now, following my 63rd birthday, I have decided to reveal all my "little secrets" here in this book as a legacy.

Mario Markus
www.mariomarkus.com

Commentary

It is interesting to inquire why these pictures from Dortmund arise such pleasure and interest. The play with graphics and shapes goes back to the very origin of mankind. It is one of the great testimonies to human evolution over thousands of centuries to see how artifacts produced by man have become progressively both more functional and aesthetically more appealing. Art and science have never been divorced.

In the rock paintings of Spain, we find numerous representations of concentric circles and spirals; curiously, on many *pis* (the ritual jades of neolithic China representing heaven), we again find engraved hexagons and spirals — precisely the most prevalent nonequilibrium "dissipative" structures we discover today in physics and chemistry.

This exhibition marks an encounter. Science rediscovers man's dreams, and art, as conceived by Kandinsky, Klee and Rothko, describes a cosmology in the making, a cosmology in which matter has encapsulated time.

Ilya Prigogine
Nobel Prize for Chemistry 1977

(This commentary is an exact reproduction of a text written in the catalogue of a previous exhibition of diagrams in this book — along with photographs of chemical waves — in Houston, Cambridge, Berlin, Dortmund and Lisbon [1][2]).

Contents

Chapter 1

The Useful and the Beautiful

Any reader browsing this book should keep this in mind: that all illustrations have a mathematical significance. Moreover, some of the underlying equations are applicable in physics, chemistry, biology or the social sciences (Chapter 8). Other equations are of purely mathematical interest and possess properties that are of general significance, i.e. they are not dependent on the properties of a specific application (Chapter 9).

I will explain the technical details in Chapter 6. In all cases, the coordinates (abscissas and ordinates) correspond to parameters that control a given system (such as the amplitude and the period of electric pulses in Sec. 8.1, the light's phase shift and the mirrors' absorbance in Sec. 8.11, or the reproduction rate of predators and prey in Sec. 8.16). At a given point on the plane, i.e. at a pair of coordinates, these control parameters are set to be constant in time, while the so-called phase variables (called x_n and y_n in this work) may change in time. In Sec. 8.16, for example, the phase variables are the relative numbers of predators and prey. The shade of grey or the colour at a point on the plane indicates how the phase variables change in time, telling us if they change in a predictable or a chaotic way. The shade of grey or colour also tells us, if the system is predictable and how fast it recovers from perturbations; if the system is chaotic, meaning that its predictability is limited like the weather, to what extent can we make finite-time predictions.

Here I want to insert a clarifying note for the non-scientist: the diagrams in this book are comparable to maps in geography. For a given point on the plane, the colours in geographical maps tell us where there is a sea and how deep it is, or and where is land and how high it is. Mirroring the function of the longitude and latitude of geographical maps, the coordinates in this book can signify quantities that characterize a system and do not change in time (such as the interest rate of a long-term bank deposit in an economical model). In an analysis of those quantities that do change in time (such as the amount of money on the bank account), the colours (or shades of grey) indicate (instead of sea or land), whether the system is predictable, and its stability when disturbed, or the extent to which it is unpredictable.

In addition to their scientific significance, the diagrams are aesthetically appealing. Note, however, that the fact that mathematical or scientific objects can stimulate us aesthetically is not a new phenomenon: simply consider the golden mean, the platonic solids and other polyedra, mazes, labyrinths, kaleidoscopic images [3] and especially fractals [4]. Moreover, there are living beings, such as radiolaria (see the impressive collection published by Ernst Haeckel [5]), as well as astronomical nebulae, which are a visual delight.

In part, this book contains diagrams which may be scientifically useful, but were nevertheless generated only with an aesthetic motivation. For example, this is the case for most diagrams in Sections 9.20, 9.21 and 9.22. However, the rules of mathematics were never broken.

Chapter 2

The "*Objet Trouvé*" in Mathematics

In 1917, the well-known artist Marcel Duchamp was so deeply inspired by a discarded urinal that he took it away, mounted on a pedestal and exposed as art. After he had done so, he became known as the initiator of "found art", which involves *objets trouvés* ("found objects").

What happened there? The philosopher Juliette Kennedy writes: "...this might at the end turn out to be a special case of the 'resonance theory' of beauty...objects are judged by us to be, say, beautiful, because properties of the object resonate with some structure within the viewer." [6]

As far as this book is concerned there are some equations (such as those of the discontinuous $\sin^2$-maps in Section 9.21) that provide an overwhelming variety of surprisingly different diagrams. This is accomplished either by changing parameters or by succesive enlargements of parts of the plane. I became addicted to this parameter searching, looking for new, aesthetically stimulating shapes within a jungle of generated outputs. The process using these equations proved to be far more fruitful than using the Mandelbrot set (see [4]), in which the shapes tend to become repetitive as the search goes on.

I am convinced that a choice among infinite possibilities is different to what Richard Wright [7] criticises as "the forgeries of computer automated art". Although the artist does not create the object in the diagrams of this book, one is actually dealing with *objets trouvés* and with the wonderful phenomenon of "resonance" described above. I believe that the situation is comparable to that of a fine art photographer who wins a prize for the portrait of an old lady in Ouagadougou. He did not create the lady, just as you or I do not "create" the equations in this book. However, the photographer performed a choice among infinite possibilities: he chose the site, the lady, the time, the angle, the depth of focus...until something happened in his mind, something that was related to his subjective world and, astonishingly, also related to the emotions of many viewers of his picture.

Coming back to this book: just as the photographer, who was faithful to the lady (in the sense that he didn't invent details or distort the photography) all diagrams in this book are faithful to the equations behind them. In spite of that, diagrams from an early part of my work [1, 2, 8, 9] won the annual exhibition prize (1988) given in Chile by the Society of Art Critics. In addition, a number of poets, including Hans-Magnus Enzensberger and Günther Kunert wrote lyrical texts inspired by my diagrams in an experiment carried out by the publishing house Birkhäuser-Verlag in Basilea. This publisher subsequently edited the book *Verknüpfungen* [10], which contains diagrams and text side by side.

Chapter 3

The Mondrian Experiments

In 1966, Michael Noll, an electrical engineer at Bell Labs in Murray Hill, published an article which is often cited in the realm of computer graphics: "Human or Machine: A Subjective Comparison of Piet Mondrian's 'Composition with Lines' and Computer-Generated Pictures" [11].

Piet Mondrian (1872–1944) had, by 1914, modified his painting style so that his pictures were composed of horizontal and vertical lines. Michael Noll created an algorithm mimicking the statistics of spacings of Mondrian's lines ("Mondrian-like pictures"). In addition, Noll created pictures with computer-generated spacings that diverged from the statistical features of Mondrian's works ("divergently spaced pictures"). He evaluated the reactions of Bell Telephone employees, as well as of people with a formal artistic training, to the different pictures. Later on, Noll himself and several other researchers [12–16] continued the investigations on original Mondrian pictures versus computer-generated ones. Some of the most notable results are listed below:

(1) On being shown a Mondrian-like picture and asked to choose whether it was "original" or "computer-generated", 72% of the participants believed it was original [11].
(2) When shown both an original and a divergently spaced picture, 59% preferred the latter [11].
(3) The results given under (1) and (2) were similar for participants with a formal artistic training and for those without [12].
(4) 57% of participants preferred Mondrian-like pictures to divergently spaced pictures [15, 16]. These results were independent of personality variables or of familiarity with Mondrian's work [15].

(5) Most participants did not prefer a Mondrian-like picture to a divergently spaced picture that was preselected as aesthetically pleasing by other participants [16].

All in all, one can conclude that the border line between man-made art and of computer-generated art in the eye of the beholder is — at least in this case — meandering. It seems that statements about the origins and the preference for one type or the other, are subjective. However, one should bear in mind that the experiments listed above include imitations of Mondrian's ideas. Contrarily, the diagrams in the present book have a different quality: it is not the computer that imitates the inspiration of an artist, as in Noll's work, but it is the "artist" inside the observer, as well as inside the author (as "object finder"), who is inspired by computer-generated pictures.

Chapter 4

An Anecdotal Report on Chaos

4.1 Twenty years ago

Twenty years ago, while working at the Max-Planck-Institut in Dortmund, Germany, I became interested in biorhythms: the endogenous, physiological clocks of organisms. I was fascinated by experiments performed in the 1960s on people in isolation that showed that our daily clock is not 24 hours, but can range between 21 and 27 hours (with some exceptional cases of up to 48 hours) and is only maintained 24 hours through exposure to light and society [17].

At that time, I asked myself what would happen to a 21-hour-person on a 30-hour-planet. Would this person adjust, and, if not: what else could happen? I decided to experiment, but instead of people I used brewers yeast cells, which have a biorhythm period of about one minute. Instead of a planet with a different rhythm, I used a motorized syringe that periodically provided the yeast with a feeding solution.

Yeast has lived symbiotically with humans for thousands of years, producing the alcohol we consume when we provide it with plants containing sugar. What most people don't know is that in many cases the production of alcohol by the yeast occurs in a rhythmic cycle.

Depending on the amplitude and frequency with which I fed the yeast, the experiment yielded three types of results: (a) the yeast's alcohol production periods became synchronised with the syringe feeding period as long as it did not vary greatly from the yeast's normal rhythm; (b) the resulting oscillations were simply the sum of the yeast's endogeneous rhythm and the feeding rhythm; and (c) chaos, i.e. the yeast's alcohol production oscillated unpredictably. My colleagues would bet me one deutsche mark that alcohol production would rise above average within the next 5 minutes. Sometimes I won, sometimes I lost.

At that time, it was the first laboratory demonstration of chaos in a biological system [18]. Nobel Prizewinner Ilya Prigogine was so fascinated by our experiment that he communicated our results to the National Academy of Sciences in the U.S., which published them in their Proceedings [19]. On the basis of this work, the University of Dortmund conferred on me the "Habilitation", a kind of assistant professorship, in 1988.

A few years later I learned that my work represented the peak of banality: I had discovered a basic phenomenon in non-linear science using an unnecessarily complicated and particular system. In fact, chaos appears in general when two oscillating systems (or oscillators) interact, provided the equations governing them are sufficiently non-linear. Many observations of chaos in such coupled oscillators were reported in those years, including the double pendulum [20] and Saturn's rings. With regards to the latter, a rock in the ring is subject to both its own endogenous rotation around Saturn (the period of which is defined by the distance) and to an external gravitational force imposed on it by the Saturn satellite Mimas. Like yeast, the rock's dynamics may become chaotic, leaving black, empty stripes at the orbits they would have if Mimas were not there.

4.2 Vertical pluralism

Before my academic fiasco became known in the university, I received a phone call from philosophy professor Friedrich Rapp in 1990, who asked me if I would like to give a talk on my work on chaos in the "Studium Generale" series in the auditorium of the University of Dortmund, Germany. Since only four speakers are invited each semester, I felt honoured and did not hesitate to agree.

Two weeks later, I saw the posters announcing my talk in the cafeteria, but they sent a shiver of fear through me: they announced a "Studium Generale" on postmodernism and I only vaguely knew what postmodernism was. I phoned Rapp, who merely said: "Take it as a challenge to find out the parallels between your results and postmodernism." Actually, I got the impression that he did not know the parallels himself — he just had some intuition. So I looked for the definition of "postmodernism" in a new philosophy encyclopedia (there was no Internet at that time). However, that text — unreadable to a layman in philosophy like me — jumped between

Heidegger, Lyotard and Bonito Oliva. I understood nothing. Luckily, it cited a work that promised to be fundamental and understandable to me. It was Leslie Fiedler's "Cross the Border — Close the Gap", which had appeared in Dec. 1969 in the magazine *Playboy* [21]. My next question was: "How do I get a *Playboy* from 1969?" Searching for an answer, I went to our librarian, who — setting aside any discomfort she may have had — looked in her files with professional countenance. To my surprise she found that there actually exists an archive of this magazine, and it was — contrary to the usual prejudices about Prussian standards — in no less than the *Staatsbibliothek Preussischer Kulturbesitz* in Berlin. Next question: "Whom do I know in Berlin?" Of course, my mother — eighty years old at that time. I phoned her, she took a taxi to the Prussian library and faxed me the article the next day.

I was amazed by the courage and the intellectual solidity of this article by Leslie Fiedler. The article — among other things — moves the focus from elitist writers like Marcel Proust and Thomas Mann, and pays tribute to songwriters like Leonard Cohen, Frank Zappa and Bob Dylan. The article's point is that some of the song lyrics of these artists reach both the literary elite and the masses — a typical postmodern feature, I learned — and therefore bridge the old abyss between an art for the educated and a sub-art for the non-educated. Further reading taught me that there exist such things as postmodern cooking and postmodern tourism, but I found nothing on postmodern science, the subject I was supposed to talk about. This was until I realized that what Fiedler was describing had been happening for years in my department, where high school science students worked on chaos theory for their training during vacations: they sometimes got valuable scientific results within a short time in spite of their lack of expertise.

This type of bridging between experts and laymen is particularly easy in chaos research both because most of this research takes place via patient, game-like experimentation at the computer and because a substantial part of chaos theory deals with the mesocosmos, i.e. the things that are directly detectable by our senses (such as lasers, flames or arrhythmic hearts). These things are certainly more comprehensible than the elitist microcosmos of the elementary particle physicists or the macrocosmos of the cosmologists.

4.3 Horizontal pluralism

The above considerations led me to consider the vertical pluralism of the elite and the masses as a postmodern aspect of chaos theory. Later on, I felt that this vertical aspect should be complemented by another "crossing the border and closing the gap": namely a horizontal, interdisciplinary pluralism, which we can associate with Bonito Oliva's "all territories of culture". This da Vinci-style pluralism had disintegrated more and more over the centuries; however, in non-linear science (to which chaos theory belongs) we witness a new convergence. Nowadays, many conferences feature such diverse experts as astronomers and medical doctors, speaking the same scientific language (see e.g. [22–24]). It is a holistic, rather than a reductionist, attitude that makes such a dialogue possible. This was exemplified above, as I illustrated the emergence of chaos in coupled oscillators. In fact, a holistic approach was demonstrated there in the choice of examples from a range of fields: classical mechanics (the double pendulum), astronomy (Saturn's rings) and biology (yeast).

Another example of horizontal pluralism is seen in the simulation of patterns in the visual cortex under near-death conditions [25]. My student Hans Schepers and I carried out these computer simulations considering so-called active sites in the visual cortex (shown in black in the outer regions and in red in the central regions of Plate 1 and on the cover of this book), which emerge erratically, are amplified by positive feedback on themselves, and activate nearby sites. In case of low oxygen supply occurring in a near-death situation, the active sites inhibit regions farther than a certain distance, preventing a further spread of activity. These inhibited regions are shown in red in the outer regions and in yellow in the central region of Plate 1. Further erratic fluctuations of activity can be amplified only in regions that are sufficiently far from the inhibitory action (so-called "lateral inhibition"). We achieved the pattern on Plate 1 when we considered that connections between the cortex and the retina are such that cartesian coordinates in the visual cortex (where the patterns are formed) are transformed into polar coordinates in the retina. This is the so-called retinocortical map. It is the corresponding retinal pattern that is interpreted as a vision by the brain. How does this relate to horizontal pluralism? The mechanism of the retina is governed by the same principles that define leaf veins or pigmentation patterns on animal skins (Turing patterns, [26]), except that activation and inhibition in leaves and skins is caused by diffusing chemicals, while patterns in the visual cortex are caused by activating and inhibiting neuronal synapses. I thus felt myself caught up in a "theoretical theology", in which we interpreted the often mentioned "light at the end of the tunnel" as something like mackerel or zebra stripes in the brain.

The spirit of horizontal pluralism is also present in the study of fractals. Fractals look similar in small and large scales, as seen in many diagrams of this book (see Chapter 10), or — most clearly — in Jonathan Swift's fleas, whose own fleas are itched by their own flea-bitten fleas and so on [27]. Fractals are studied by employing the same methods used to study the surface of catalysts, as well as reconstructed attractors from the sun's activity or from EEG-signals recorded during different states of the mind. Horizontal pluralism via holism is also the essence of the theory of excitable media [23, 28, 29], which embraces systems as diverse as heart muscle, epidemics, aggregating slime mold, fertilized fish eggs and galaxy dynamics.

4.4 Surprises contrary to reason

I define a third postmodern aspect of chaos theory as "openness to surprises contrary to reason": the celebration of arbitrariness...reason as the Goddess of Modernism...and of the Age of Enlightment. By using reason, humans have made projects for their future through various ideological programs: the Reformation, the Counter Reformation, German Idealism, Marxism and, in the natural sciences, Newtonism. I use the word "reason" here in the sense of "calculability". Newton's work aroused a claim for predictability that peaked with Pierre Simon de Laplace in 1776. Laplace envisaged an entity, later called the "Laplace Demon" (nobody envisaged anything like a computer at that time), capable of calculating the past and future of the universe, provided we feed it with the positions and velocities of all particles of the universe, and, of course, with Newton's formulae. It follows from Laplace's thinking that if we were to include the particles in our brains, free will would become an illusion. Certainly Laplace's God is equivalent to the present state of the universe since it is the same thing we call "past", then "present" and then "future", because one may calculate one from the other. Indeed, Laplace's God simply turns the pages of a book that is already written.

Contrarily, postmodernism abandons the idea of an historically structured and predictable world *à la* Isaac Newton or Karl Marx. It is characterized by the acceptance of arbitrariness. Strangely, scientists have stubbornly stuck to calculability, although there are phenomena visible to everyone that have always resisted prediction: weather changes, the roulette wheel and sudden insect plagues such as the forest-eating gipsy moth in New England: people think it's snowing in the wrong season, but it's moths that they see.

The first scientist who realized that Newton's equations may lead to unpredictable behaviour was Henri Poincaré. He studied, at the end of the 19th century, the problem of three interacting celestial bodies (such as in the example given earlier: Saturn, a ring's rock and Mimas) and showed that small perturbations in the initial conditions of such a system may eventually overshadow the whole system. This was later called the "butterfly effect". Poincaré remarked: "Unthinkably small subtleties in the croupier's arm may decide my stake in the casino.... These things are so bizarre that I cannot stand thinking about them" [30]. This was the birth of what was later called the science of chaos. However, this science stayed dormant for about 70 years because most calculations require computers. A revival occurred in the 1960s when the meterologist Edward Lorenz could solve his equations [31] on a computer. Lorenz's equations are given in Section 7.3.

With a potentially chaotic system, such as oscillating yeast, a mechanical device or an insect population, it is important to know which externally imposed conditions make the system behave chaotically and which render it predictable. This is the scientific motivation of the diagrams in this book. As I mentioned in Chapter 2, a collection of these diagrams garnered an award in Chile — not from scientists, but by a society of art critics. That surprising event extends horizontal pluralism beyond its status as a postmodern aspect of chaos theory into an even broader realm in which pictorial art is connected to scientific activities: C.P. Snow's "two cultures" [32] merge.

Also relevant is the concept of the "controlled accident", which has led to the comparison of computer-generated pictures and the paintings of Jackson Pollock. As Timothy Binkley noted, "The computer rises from the sea of postmodern culture...as a wily sorcerer...that commands an intriguing repertoire of artistic resources" [33]. A more elementary and more critically examined analogy is the kaleidoscope, which David Brewster exhuberantly presented as a source of art in the nineteenth century [3].

The overwhelming development of computers generated a boom in chaos studies. Unfortunately, many of its results are only of dubious use (for example, the interpretation of the outbreaks of some epidemics as phenomena related to deterministic chaos [34]). On the other hand, some highly useful results appeared in the clinical domain. One example is Rolf-Dieter Hesch's finding (at the Medizinische Hochschule Hannover, Germany) that parathormone oscillations of healthy people are chaotic and that these oscillations become more and more predictable as osteoporosis develops in a patient [35]. There are more interesting examples relating chaos to medicine, including electrocardiogram analyses and the development of implantable defibrillators to control chaotic heart rhythms.

Considering all these efforts: what have we learned about the predictability of conditions like the weather? The routine is to collect the data from about 10,000 weather stations (plus some satellites and balloons) and get predictions for about three days. Chaos theory allows us to estimate that a prediction 11 days in advance would require 100 million stations, and a prediction one month in advance, 10^{20} stations all over the world — that is, one station on every 5 mm^2 of earth and water. Clearly, there is no feasible solution, and the Laplace Demon is in agony. Even more aggravating is the estimation that in order to process the data for a two-month prediction, one would need a computer that has more processing elements than there are atoms on earth. Furthermore, no presently conceivable computer speed would render any results before the two months were over.

We can definitively kill the Laplace Demon (more precisely, the meso- and macroscopic Demon, since quantum mechanics has already killed the Demon at the atomic level). We can kill it with a more lucid example than the weather: that of balls colliding on a billiard table. Humans can learn to predict the last angle of a 3-ball carom. Any advanced physics student can calculate how amazingly quickly unpredictability increases with the number of balls. A computer programmed to predict the outcome with 17 balls would have to consider the gravitational forces of objects around the table. For 49 balls (the number of balls in the German lottery sphere), the gravitational force of the whole Milky Way is involved. For 56 balls one would have to consider the influence of an electron 20 billion light years away, at what some would call the "edge of the universe".

Undoubtedly, a computer exists that is able to consider all these influences. It is an analogue computer: the universe itself. Only the universe can predict what will happen within it. As a philosophical layman, I am deeply amazed by the development of our concept of the world and of its knowledge. In 1586, Giordano Bruno published his *Figuratio Aristotelici Physici Auditus*, in which he rebelled against the predominance of Aristotelianism and wrote that God is the universe, which drives itself. The main difference between this stance and my statement above on the self-predicting universe is that Bruno was burned for his position. In 1687 (101 years later), Isaac Newton — free from Aristotle's ideas — published his *Philosophia Naturalis Principia Mathematica*, the seed of what would become Newtonism. Another 101 years later, in 1788, Newtonian science reached its peak with the publication of Joseph-Louis Lagrange's *Mécanique Analytique*, contemporaneously with Laplace's Demon. In 1889 (yet another 101 years later), our view of the world changed with Poincaré's work on the three-body problem, *Les Méthodes Nouvelles de la Mécanique Céleste*, the seed of chaos theory. And still another 101 years later, in 1990, I received a phone call from Friederich Rapp, which, via *Playboy* magazine (see Section 4.2) led my worldview to encounter that of Giordano Bruno. Thus, a big circle, embracing centuries of knowledge, was closed. (Note: I owe the idea of the 101-year cycle to my friend Professor Peter Richter of the University of Bremen.)

One must keep in mind, however, that the world is not only made of unpredictable phenomena. Technical devices or biological functions that make the world reliable operate in the regions of predictability (pictured mainly as the lighter regions in the foregrounds of this book's diagrams). On the other hand, we know that such wonderful things as orange trees and the human brain resulted via selection from unpredictable fluctuations. Similar scenarios are found in the development of ideas. Hans Jensen, who taught me theoretical physics in Heidelberg, enjoyed recounting how he got the idea for his nuclear shell model: watching dancers in a carnival after he had been brooding fruitlessly over huge amounts of spectroscopic data. Some say he was drunk. In any case he got the Nobel prize for the idea.

Although the interplay of unpredictability (as creative drive) with predictability (for functional reliability) has indeed been recognized by scientists (especially evolutionary biologists), I am not sure that sociologists have learned from this dissolution of a black and white mentality. In fact, the modernist or postmodernist paradigm has been subject to serious political discussions. A. Hill made his position clear in 1987 [36]: "Modernism was largely Left, postmodernism...is largely Right, if not covertly Fascist". Note that postmodernism became exceptionally fashionable around 1990, after the dissolution of the Eastern bloc. When Russia then suffered a severe crisis that affected the rest of the world, Marxist Terry Eagleton achieved a great deal of publicity with his book, *The Postmodern Illusion* [37], which induced newspaper headlines such as *The Postmodern Game is Over* and *The Postmodern Swindle.* Thereafter, the tensions dissolved.

The reader may be expecting from me some formula that would instigate in politics a fruitful interplay between planned and unplanned economic models. However, being a scientist, I will not dare to suggest operating instructions. Instead, I can offer a phenomenological description of such an interplay by using the recent theory of "Class 4" behaviour (S. Wolfram [38]). This is also called "the edge of chaos". "Class 1" is stationarity, stagnation; "Class 2" is periodicity; "Class 3" is chaos; and "Class 4" is the unpredictable alternation of chaos and order. There exist, indeed, systems that endogeneously display "Class 4" behaviour, which has been regarded as essential for the development and survival of living systems (see the books by M. M. Waldrop and R. Lewin [39, 40]). However, no matter to what degree we accept a general relevance of Class 4 systems, the problem remains: the predictability-unpredictability transitions are again unpredictable. Thus, we may be closer to understanding what is going on, but farther away from knowing what to do next. Therefore, any attempt at a modern-postmodern parliament will probably itself be postmodern.

Plate 1 represents an example of self-organization in the human brain cortex. Being the simulation of a "vision", this figure also puts us closer to the mystical wandering through the last tunnel in *The Rise to the Empyreum* by Hieronymus Bosch. To be honest, though, I suspect myself of having deliberately influenced the choice of parameters for Plate 1 so that it would capture the atmosphere of Bosch's masterpiece, or maybe of Doré's illustrations of Dante's paradise-vision. I certainly did so in Fig. 9.136, having other works of Bosch, namely parts of *The Gardens of Lust* (see [41]) in mind; since this figure also transmits a scientific message, I once more felt deeply involved in a dissolution of the concept of the "two cultures".

Chapter 5

A Case Submitted to Court

Many of the pictures in this book show tiny light regions ($\lambda < 0$) within a black backround, or *vice versa.* (Examples are Figs. 8.19, 8.22, 9.2, 9.19, 9.55, 9.63 and 9.73). Considering that the lighter regions stand for predictability (periodicity) and that black means unpredictability (chaos), we are faced with a dramatic phenomenon: very small changes in the parameters that control a system (corresponding here to the coordinates) cause drastic changes in the dynamics (see also Plates 13 and 26). In such cases, one speaks of "structural instability". This concept is not to be confused with the so-called "butterfly-effect", which appears in unpredictable (chaotic) systems and tells us that for fixed parameter values, small differences in the initial conditions can lead to very different outcomes (Section 4.4). Note that structural instability can be more dramatic than the "butterfly effect" because, given small errors in the parameters, one cannot even state that the system is unpredictable, as it may be predictable as well.

A physical problem that has kept physicists busy for more than a century is that of a planet orbiting around two equal, nearby, suns. It was only after the development of efficient computers that long-term trajectories were calculated for that system. A group at the University of Bremen, Germany (see [4]) showed that, depending on the conditions, the planet may move periodically or chaotically. However, in 1993 another group, working at the University of Karlsruhe, Germany, showed that the trajectory of the planet is always periodical. For the conditions they chose, the scientists in Karlsruhe showed that if the time steps for numerical integration are sufficiently large, the calculated trajectory is chaotic, while at smaller time steps it is periodical; they boldly deduced from this that chaos is a numerical artifact. The popular weekly magazine *Der Spiegel* [42] interviewed the fellows from Karlsruhe and declared in a sensationalist article that chaos is a big lie, since the results depend on the way the problem is handled by the computer. *Der Spiegel* went even further, stating that the whole body of research on chaos is nonsense: dozens of professors from different universities were discredited. As to myself, they visited me without revealing their destructive intentions and then they published one of my diagrams with the caption: "No more significant than wallpaper design".

The scientists in Bremen, who were offended by being called liars in a widely read journal (see [43]), took the case to court. They could prove that the finding of their opponents in Karlsruhe depended on the particular choice of parameters. It can, in fact, be shown that structural instability holds. "Justice" was done by enacting a counterstatement with so little substance that it didn't attract much attention. However, every scientist who knows the work of Poincaré, Arnold, Smale, Ruelle and the strict proofs of Sinai for the existence of chaos, may have just smirked or ignored the whole circus.

As to the present book, one should keep in mind that results may be sensitive, not only to the parameters in the equations and to the initial conditions, but also to an important numerical parameter: the time of integration (the number of iteration steps). However, I do hope that if a reader finds divergent results here and there when iterating with more steps, I will not have to meet that person in court.

Chapter 6

Calculation of the "Charts for Prediction and Chance" (λ-Diagrams)

Note: An informal glossary can be found in Appendix A and abbreviations are given in Appendix B.

In the following one finds the mathematical method used to generate the diagrams in this book. This is also the method used in the attached CD-ROM. Only one-dimensional (1D) and two-dimensional (2D) maps (with discrete time steps $n = 0, 1, 2, ...$) are considered. (Their scientific significance will be discussed in the next Chapter 7.) Thus, equations of the type

$$x_{n+1} = f(x_n) \tag{6.1}$$

(1D), or

$$x_{n+1} = f(x_n, y_n) \tag{6.2}$$

$$y_{n+1} = g(x_n, y_n) \tag{6.3}$$

(2D) are iterated, starting at an initial value x_0 or (x_0, y_0), which is supplied by the user. A plausible guess, which depends on the system, must be made for this initial condition. Note that the results may depend on the condition chosen.

For both 1D and 2D maps, the coordinates in the diagrams are parameters appearing in f or g. Thus, each point on the plane corresponds to a pair of parameters. The colour or the shade of grey at a given point corresponds to the maximum Lyapunov exponent λ, which will be explained in the following section.

6.1 The Lyapunov exponent

A general analysis shows that after a time t (t large) a small perturbation in a continuous system is multiplied by a factor $e^{\tilde{\lambda} t}$. For a discrete map, such a perturbation is multiplied by a factor $e^{\lambda n}$. If $\lambda > 0$, then the perturbation grows. This is the definition of chaos and the corresponding growth of a perturbation is the famous "butterfly effect". However, the smaller λ is, the better short-term predictions will be: therefore, limited predictability can be quantified by the magnitude of λ.

Before continuing on with the case $\lambda < 0$, I would like to insert an hilarious note. Long before chaos was associated (in the 1960s) with a butterfly in India causing a tornado in Texas, W.S. Franklin wrote in 1898 [44] that "Long range detailed weather prediction is impossible...the accuracy of this prediction is subject to the condition that the flight of a grasshopper in Montana may turn a storm aside from Philadelphia to New York". This statement was certainly inspired by the work of Poincaré and has the folkloric implication that there exist pedants who would like to rename the phenomenon to the "grasshopper effect".

If $\lambda < 0$, then any perturbation is damped, so that we are dealing with fixed points, or periodical (and thus predictable) phenomena. The value of $|\lambda|$ for $\lambda < 0$ tells us how fast the system recovers from a perturbation, i.e. how stable the system is. If one is dealing with a biorhythm or building a cyclic machine, then, of course, one wishes for negative λ with large $|\lambda|$.

For 1D maps a "perturbation" dx_i grows to dx_{i+1} in the next step. Writing $|dx_{i+1}| = |dx_i| e^{\lambda_i}$, where λ_i is the contribution to λ in the time step i, one obtains $\lambda_i = \ln \left| \frac{dx_{i+1}}{dx_i} \right|$. Thus, λ is given by the average

$$\lambda = \lim_{N \to \infty} \frac{1}{N} \sum_{i=1}^{N} \ln \left| \frac{dx_{i+1}}{dx_i} \right| . \tag{6.4}$$

Therefore, given a 1D map

$$x_{n+1} = f(x_n) \tag{6.5}$$

all one has to do is to calculate the derivative $f'(x_n) = \frac{dx_{n+1}}{dx_n}$, take the logarithm of its absolute value and average over a large number of steps.

Let us now consider 2D maps

$$x_{n+1} = f(x_n, y_n) , \tag{6.6}$$

$$y_{n+1} = g(x_n, y_n) . \tag{6.7}$$

The x- and y-components of a perturbation define a vector

$$\delta\vec{x}_n = \begin{pmatrix} \delta x_n \\ \delta y_n \end{pmatrix} .$$

If $\delta\vec{x}_n$ is sufficiently small, one can approximate its evolution by

$$\delta\vec{x}_{n+1} = J(\vec{x}_n)\delta\vec{x}_n , \tag{6.8}$$

where

$$J(\vec{x}_n) = \begin{pmatrix} \frac{\delta x_{n+1}}{\delta x_n} & \frac{\delta x_{n+1}}{\delta y_n} \\ \frac{\delta y_{n+1}}{\delta x_n} & \frac{\delta y_{n+1}}{\delta y_n} \end{pmatrix} \tag{6.9}$$

is the Jacobian matrix, evaluated at $\vec{x}_n$. In the present book, the initial perturbation was set to

$$\delta\vec{x}_0 = \begin{pmatrix} \sqrt{2} \\ \sqrt{2} \end{pmatrix} . \tag{6.10}$$

It was multiplied by $J(\vec{x}_0)$, yielding the perturbation $\delta\vec{x}_1$ of $\vec{x}_1$. Then, $\delta\vec{x}_1$ was normalized to length 1 (dividing by $d_1 = |\delta\vec{x}_1|$), then multiplied by $J(\vec{x}_1)$, yielding $\delta\vec{x}_2$. Proceeding as before, $\delta\vec{x}_2$ was normalized to length 1 (dividing by $d_2 = |\delta\vec{x}_2|$), and so on. Finally, λ was determined by

$$\lambda = \frac{1}{N}\sum_{i=1}^{N} \ln d_i . \tag{6.11}$$

The reason for the normalization to length 1 at each step is the prevention of overflows (or underflows), since the perturbation changes exponentially. An important note is that there are actually two independent directions $\vec{\epsilon}_1$ and $\vec{\epsilon}_2$ (eigenvectors) in the plane, changing by virtue of the eigenvalues λ_1 and λ_2 of the Jacobi matrix. Accordingly, there are two Lyapunov exponents, so that a perturbation $\delta\vec{x}_0$ changes as

$$\delta\vec{x}_n = \delta x_0\, e^{\lambda_1 n}\vec{\epsilon}_1 + \delta y_0\, e^{\lambda_2 n}\vec{\epsilon}_2 . \tag{6.12}$$

If $\lambda_1 > \lambda_2$, then one can write

$$\delta\vec{x}_n = \delta x_0\, e^{\lambda_1 n}\left(\vec{\epsilon}_1 + \frac{\delta y_0}{\delta x_0} e^{(\lambda_2 - \lambda_1)n}\vec{\epsilon}_2\right) . \tag{6.13}$$

The second term in the bracket will obviously tend to zero as $n \to \infty$, so that only the largest Lyapunov exponent will become noticeable in the long-term development of a perturbation. In other words, the value of λ obtained by the iterative procedure described above is actually the maximum Lyapunov exponent λ_1. It is this exponent that is indicated by colours or shades of grey on the diagrams of this book.

Before λ is calculated, for both 1D and 2D maps, as described above, a number of iteration steps has to be performed in order to allow transients to die away. We call them "pre-iterations" and indicate them by n_{prev} in the figure captions. As a rule of thumb, a few hundred of them are usually done, but there are cases in which 20 are enough and others in which 2000 are necessary. In scientific work, one should vary this number and observe whether or not λ changes significantly. The pre-iterations are not included in the calculation of λ because the initial values x_0 in 1D or (x_0, y_0) in 2D are arbitrarily set by the user and thus do not necessarily belong to the set that is finally reached by the system. Besides n_{prev}, another quantity indicated in the figure caption is n_{max}. The total number of calculated iterations is $n_{\text{prev}} + n_{\text{max}}$, but only n_{max} is used to determine λ.

Furthermore, any step in which the matrix $J(\vec{x}_i)$ is equal to zero (in 2D) is ignored in the summation of the logarithms, since such a matrix leads to a perturbation equal to zero and normalization fails. The transformed perturbation $\delta\vec{x}_{i+1}$ is then reset to the initial value $\delta\vec{x}_0$. This resetting does not affect the iterations (using the functions f and g): they are performed regardless of whether $J(\vec{x}_i)$ has been equal to zero or not. An example for a case in which the Jacobian matrix is equal to zero is given in Section 9.18 (Mandelbrot set) at the very start of iterations, since $(x_0, y_0) = (0, 0)$.

6.2 The use of colours and shades of grey

Many criteria were used for the assignment of colours in this book, depending on which features one wanted to highlight. However, in most pictures a colour discontinuity at $\lambda = 0$ was used in order to indicate the transition from predictability to chaos. (This is comparable to the transition from earth to sea in a geographical map.) Within the regions with $\lambda < 0$, colours were changed either in several discontinuous steps (e.g. in Plates 11 and 40) or continuously (e.g. in Plates 9 and 10).

In black and white maps, a discontinuity from white to black was set at the transition from $\lambda < 0$ to $\lambda > 0$. Furthermore (considering grey levels 0 through 255), the grey level 1 (black) was used in this book whenever there was chaos ($\lambda > 0$). If $\lambda < 0$, the shade was changed continuously from black (lowest negative values of λ) to white (largest negative values of lambda). Whenever λ diverged, becoming smaller than 10^{-20} or larger than 10^{20}, the grey level was set to 0 (black). The remaining grey levels (1 through 255) were used for the non-diverging λs.

In some cases, which we indicate with "L-shading" (linear shading) in the captions, the interval I in the real axis between the minimum and the maximum negative λ was divided into 255 equal sub-intervals, and to each sub-interval a grey level between 1 (blackest; interval with smallest λ) and 255 (white; interval with the largest negative λ) was assigned. A given pixel was then painted with the grey level corresponding to the sub-interval containing the λ at that pixel. This was done in Figs. 8.3, 8.6, 8.7, 8.15, 8.19 and many others. If instead of setting the lower limit of the interval I by the minimum of λ, we set it to some number λ_m, then the value of λ_m is written in the caption, e.g. $\lambda_m = -2$ in Figs. 9.132 and 9.157.

In most cases, however, visualization is easier when the grey levels are distributed in a way I call "democratic". This is indicated by "D-shading" in the captions. The algorithm in this case is as follows. Suppose the number of pixels with negative λ in a diagram is S. Here, "democracy" means that S/255 pixels correspond to each of the 255 grey levels. The interval on the real axis between the smallest and the largest negative λ is divided into 10^4 equal bins in order to determine approximately the distribution $p(\lambda)$, where p is the probability of finding a pixel with Lyapunov exponent λ among the S pixels with $\lambda < 0$. As the next step, the interval between the smallest and the largest negative λ is divided into 255 subintervals $I_i = [\lambda_i, \lambda_{i+1}]$ $(i = 1, 2, ..., 255)$ satisfying

$$\int_{\lambda_i}^{\lambda_{i+1}} p(\lambda) d\lambda = S/255 \,. \tag{6.14}$$

A grey level i is assigned to each of the 255 subintervals I_i, ranging from $i = 1$ (black) for I_1 up to $i = 255$ (white) for I_{255}. Given a pixel, λ is determined in that pixel (see Section 6.1); then the interval I_i which includes that λ is determined and the pixel is painted with the grey level i. This "democratic painting" causes, as a side effect, an apparent three-dimensionality in the graphical structures within the regions of the diagrams where $\lambda < 0$.

Note that, due to the colour or grey level discontinuity at $\lambda = 0$, the pictures in general appear as if the regions with $\lambda < 0$ are objects in the picture foreground. In contrast, the regions with $\lambda > 0$ appear, in general, as the background of the picture.

6.3 Configuration of the diagrams

Parameters appear in two distinct "types" in the abscissas and ordinates in this book. In one type, two independent parameters of the maps are represented, e.g. α and T in Fig. 8.1. In the other type, all parameters are kept fixed except one, which alternates periodically between two values A and B as iterations proceed. (This is the case, for example, in Fig. 8.2.) A and B are given in the coordinates. The periodicity of A and B can have different forms: ABABAB..., AABAB AABAB..., or A^3B^3 = AAA BBB AAA BBB..., as indicated in each case. (For example, if a parameter r is varied as ABAB... then we write r:AB AB... in the figure caption.) In these latter cases, we are simulating coupled oscillators: one given by the map and one given by an alternating parameter, which by virtue of a dichotomic approximation may assume only two values, A and B.

The representation of λ on a plane defined by two control parameters ("parameter plane") is called here a "λ-diagram". Note, however, that many diagrams in this book were obtained by assigning one parameter to the x-axis, the other to the y-axis and then rotating the coordinates (usually counterclockwise by 45°), as can be deduced from the coordinates of the corners, which are given in the captions (LL: lower left corner, UL: upper left corner and LR: lower right corner).

When looking for a new, interesting λ-diagram the following steps are recommended: (1) If the equation is not the result of a scientific task and is thus not dictated by necessity, one first figures out a map, e.g. by modifying the maps in this book; (2) λ-diagrams with the largest possible ranges of coordinates are generated with varying parameter sets; (3) sections of the λ-diagrams generated in the preceding step are magnified; (4) rotation of a chosen diagram, as well as stretching or compression of coordinate-axes, is performed if appropriate. Figures 9.48, 9.55 and 9.80 provide examples for the transition from step (2) to step (3). In fact, mere inspection of the figures reveals small regions that may well be interesting after enlargement. In some cases, magnifications of successively smaller sections, yielding surprising new diagrams, can go on *ad infinitum*.

Chapter 7

The Significance of Discrete Maps

In this book, we only examine processes that are discrete in time. These processes are described by maps of the form given at the beginning of Chapter 6. On the other hand, processes in nature and elsewhere are habitually continuous. In view of this (apparent) contradiction, this chapter deals with the question: "What is the practical relevance of discrete maps?" I will show now that not all interesting processes are continuous and that continuous processes can be discretized. The individual sections analyse different cases.

7.1 Processes that are inherently discrete in time

As a first example, consider a deposit x_n in a bank with an annual interest rate z. In the years $n = 0, 1, 2, \ldots$ the deposit grows as

$$x_{n+1} = (1 + z)x_n \ . \tag{7.1}$$

(Note: the parameter z indicates an interest rate of $100z\%$.) Imagine now that a government enacts a law pronouncing that interests will only be paid to customers with a deposit below x_c, while richer investors with a deposit over x_c in value must pay a tax instead. This government calculates interest rates or taxes by

$$z = z_0 \left(1 - \frac{x_0}{x_c}\right) \ . \tag{7.2}$$

Inserting this equation for z into the equation for x_{n+1} yields

$$x_{n+1} = (1 + z_0)x_n - z_0 \frac{x_n^2}{x_c} \ . \tag{7.3}$$

It can easily be verified that by setting e.g. $x_c = 0.72$, the deposit will fluctuate chaotically for $z_0 > 2.57$. In other words, if too much is given to the poor, the bank will be more or less a casino for the rich.

A second example is the biomathematical description of the growth of leaves in a circle. Leaves appear after certain time intervals, meaning that the system is inherently discrete in time.

Leaves in a circle are found in some flatly growing succulent plants, artichokes and — if accompanied by vertical growth — in pineapples, pinecones and some thistles. I shall present here an approach that is based on the work of H.J. Scholz [45]. Let us assume a circle and a leaf that has just appeared. This leaf produces an inhibitory substance (a peptide, in general) that prevents the appearance of another leaf until this inhibitor has sufficiently decayed (so-called "lateral inhibition"). Note that I have already referred to such an inhibitory substance in Section 4.3. One can approximate the spatial distribution of the concentration of the inhibitor h_i by a Gaussian function:

$$h_i = b\, e^{-\left(\frac{x}{a}\right)^2}\ , \tag{7.4}$$

where a and b are characteristic parameters of the plant, i is the index indicating the sequence of leaf growth ($i = 0$ for the first leaf, $i = 1$ for the next, and so on) and x is the euclidean distance between a point on the border of the circle and the leaf. (See Fig. 7.1).

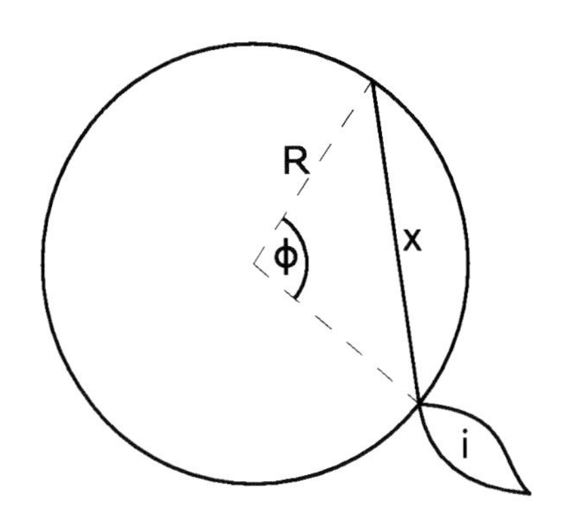

Fig. 7.1 A leaf i growing in a circle.

A convenient parameter is the angle ϕ satisfying

$$\frac{x}{2R} = \sin\frac{\phi}{2}\ , \tag{7.5}$$

where R is the radius of the circle. This allows to express h_i as a function of ϕ. After the appearance of n leaves, the total concentration of inhibitor at an arbitrary angle ϕ is

$$H_n(\phi) = \sum_{i=1}^{n} h_i(\phi - \phi_i')\ , \tag{7.6}$$

where ϕ_i' corresponds to the position of leaf i. It is assumed that $H_n(\phi)$ decays all over the circle at a rate that is independent of ϕ. A new leaf appears at time t_n at a site ϕ_n such that the total concentration of inhibitor there is below a threshold θ. (θ is another parameter that characterizes the plant).

At time t_{n-1}, the total inhibitor concentration is $H_{n-1}(\phi_n)$ at site ϕ_n and $H_{n-1}(\phi)$ anywhere else. At time t_n the inhibitor concentration is θ at site ϕ_n and $X(\phi)$ anywhere else. Since the decay of the inhibitor is independent of ϕ, one can write

$$\frac{H_{n-1}(\phi)}{H_{n-1}(\phi_n)} = \frac{X(\phi)}{\theta} . \tag{7.7}$$

Solving this equation for $X(\phi)$ and adding the inhibitor concentration $h_n(\phi)$ produced by the new leaf, one obtains

$$H_n(\phi) = \theta \frac{H_{n-1}(\phi)}{H_{n-1}(\phi_n)} + h_n(\phi) . \tag{7.8}$$

This is the discrete map we have been looking for. However, an additional step is required: ϕ_{n+1} must be determined by minimizing $H_n(\phi)$ with respect to ϕ. It is at this minimum of $H_n(\phi)$ where the threshold θ will be first reached as the inhibitor decays. Then, a new iteration can be performed using the discrete map, subsequently minimizing the total inhibitor concentration, and so on.

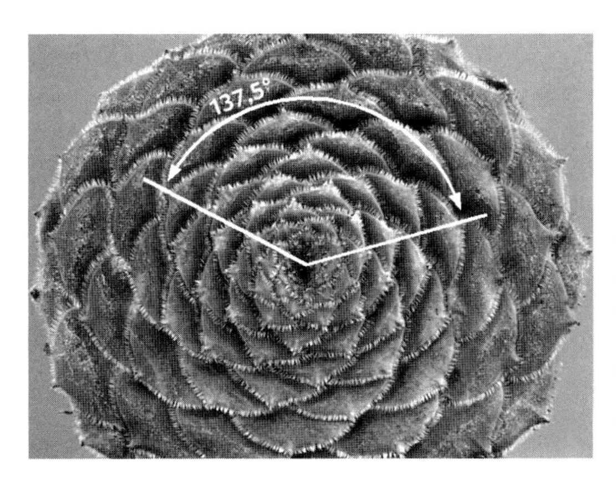

Fig. 7.2 Leaves of *Aeonium tabulaeforme* (Canary Islands), growing almost on a plane. The numbers indicate the sequence of leaf appearance. The angle between successive leaves (exemplified in the figure) is close to the "golden angle" 137.5° (Photo: W. Barthlott).

Modifying the parameters θ, $\frac{a}{R}$ and b one obtains (after allowing transients to die away) the angles $\phi_{n+1} - \phi_n = 137.5°$ or $99.5°$ for most reasonable parameter combinations. The first of these angles (the "golden angle") is related to the golden section of the circle. In fact, $137.5° \approx 360° - \frac{360°}{\tau}$, where $\tau = \frac{(1-\sqrt{5})}{2}$ is the so-called golden ratio. It has been shown that the two values obtained for $\phi_{n+1} - \phi_n$ allow the most "democratic" distribution, in the sense of minimum sunshine-blocking of one leaf by the others [46, 47]. For a small region of parameter space, $\phi_{n+1} - \phi_n = 180°$ is also obtained; this is often observed in plants with additional vertical growth. Anyway: next time you see a thistle or a plant with its leaves growing in a circle, watch out for the 137.5° and impress your date with this discrete map story. (Don't forget your protractor!) Alternatively, just look at Fig. 7.2 here. Note also that whenever the golden angle appears and the plant grows upwards, n clockwise spirals and m counter-clockwise spirals (so-called parastichies) appear in such a way that n and m are subsequent Fibonacci numbers $(1, 2, 3, 5, 8, 13, 21, ...)$, i.e. numbers obtained sequentially by summing the previous two numbers after starting with 1 and 2 [48]. (Count the spirals of pinecones in the woods and at pineapples in the grocery store and impress your date once more.) This "magic" section closes by noting that the ratio of subsequent Fibonacci numbers approaches the golden number τ, as their magnitude tends to infinity. And what about the other angle between leaves, namely 99.5°? This one yields spirals corresponding to subsequent numbers of the "small Fibonacci set", which starts with 1 and 3 instead of 1 and 2: 1,3,4,7,11,18,...

7.2 Data obtained from a continuous process at discrete times

A famous example of this case is that of Ricker, who observed salmon populations x_n (n corresponds to the year) at the Pacific coast of Canada and could fit to these observations the map

$$x_{n+1} = x_n \, e^{r(1-x_n)} \, . \tag{7.9}$$

(See [34].) A similar equation, namely

$$x_{n+1} = a \, x_n \, e^{-b x_n} \, , \tag{7.10}$$

has been fitted to the number of people infected with mumps and rubella in Copenhagen over several decades. In my opinion, however, these epidemiological fits are not convincing [34].

More convincing examples are found in physiological rhythms, as shown in Fig. 7.3, where one sees a 1D dependence (approximately) of the potential (in mV) of a neuron of the marine pulmonate mollusc *Onchidium verruculatum* [49]. The discrete times are those corresponding to a fixed phase of the periodic forcing. It shouldn't be difficult to find a function

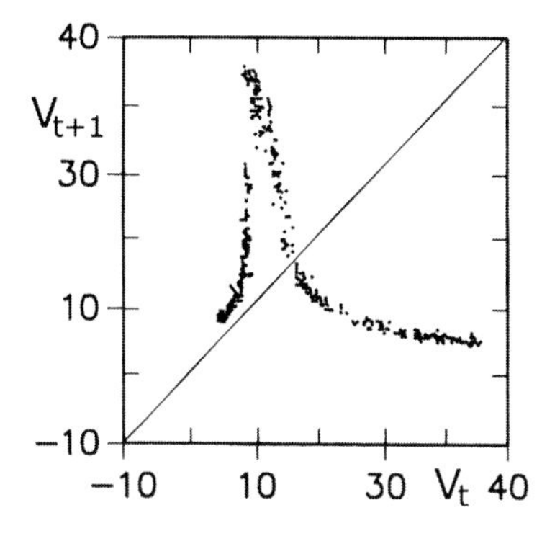

Fig. 7.3 Membrane potential of the giant neuron of the mollusc *Onchidium verruculatum* under periodical forcing. The abscissa shows the potential at a given phase; the ordinate shows the potential one forcing period later [49].

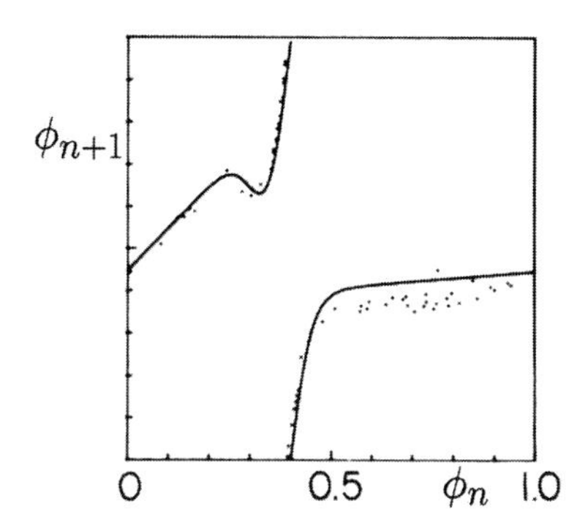

Fig. 7.4 Next-phase plot of experiments (points) and simulations (curves) of embryonic chick heart cell aggregates, which were stimulated by periodical electrical pulses. (Adapted from [50]).

fitting these data. Similar experiments were performed for periodically stimulated chick heart cells. Results are shown in Fig. 7.4 (points); a fit with a 1D map is given by the continuous curve in the figure.

7.3 Poincaré sections

A Poincaré section is defined with a cut across the phase space with a plane, as illustrated in Fig. 7.5. By virtue of such a cut, a point (x_n, y_n) is mapped into a point (x_{n+1}, y_{n+1}) and one obtains two-dimensional maps (from one passage of a trajectory through the plane to the next passage) as integrals of the differential equations describing the system. The integration of the differential equations can, in general, be done numerically. In other words, a Poincaré section and numerical integration allows us, in principle, to draw diagrams like those in this book for any physical, chemical, biological, economical or mathematical system, as long as the differential equations are written down.

In spite of these broad perspectives, in this book only explicitly written maps are considered, i.e. maps for which one can write down a formula for the functions relating (x_{n+1}, y_{n+1}) to (x_n, y_n). In other words, this book does not deal with cases for which numerical integrations have to be performed between successive Poincaré cuts. The reason for this restriction is that computing times are still dauntingly long when it comes to integrating

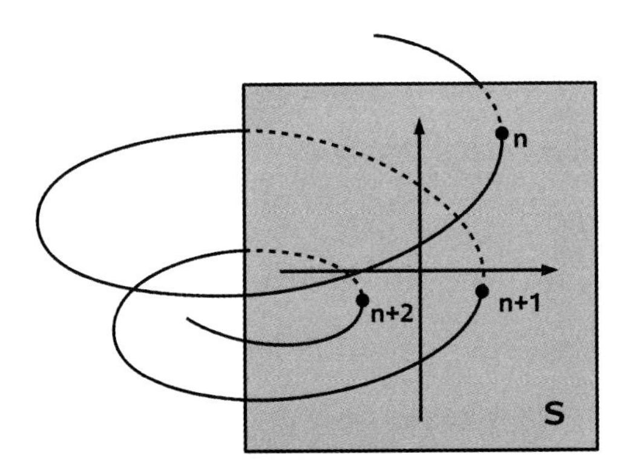

Fig. 7.5 Scheme for the construction of a Poincaré section $(x_{n+1}, y_{n+1}) = f(x_n, x_n)$ on a surface S that cuts a continuous trajectory.

over and over, not only between passages through the cutting plane, but also for widely varying parameters if a detailed analysis in parameter space is to be performed. Sometime in the future, when computers have become sufficiently fast, another book may appear involving Poincaré sections obtained by succesive numerical integrations and thus having a much larger range of applications.

In some cases integration can be avoided in the determination of 2D maps from differential equations. In fact, 2D maps can be written down as an approximation of points in a Poincaré section, as in Holmes' approximation of Duffing's differential equation (see Section 8.18) or in Jeffries' approach describing p-n-junctions (see Section 8.13). Things become especially easy if the smallest Lyapunov exponent is so negative that it substantially flattens a chaotic attractor, which can thus be considered (almost) as a surface bending in three-dimensional space. In such cases, the differential equations yield a 1D map. In fact, a Poincaré section (2D cut) of such an attractor would be a curve, albeit with some negligible width, so that it can be approximately described by a single variable. For the determination of the 1D map, this variable is discretized into steps defined by the passages of the trajectory trough the cutting plane.

An example for a quasi 2D attractor is obtained from the equations

$$\frac{dx}{dt} = -\sigma\, x + \sigma\, y \tag{7.11}$$

$$\frac{dy}{dt} = r\, x - y - x\, z \tag{7.12}$$

$$\frac{dz}{dt} = -b\, z + x\, y\ , \tag{7.13}$$

($r = 28, \sigma = 10$ and $b = 8/3$), which were solved by Edward Lorenz in the early 1960s [31]. It is a weather model that was simplified so roughly that it is of no practical use in meteorology, but the very fact that Lorenz could compute it paved the way for the chaos boom that followed. A cut through this attractor yields sets that are almost curves. These sets can be described by their x- and y-values alone, as illustrated in Fig. 7.6.

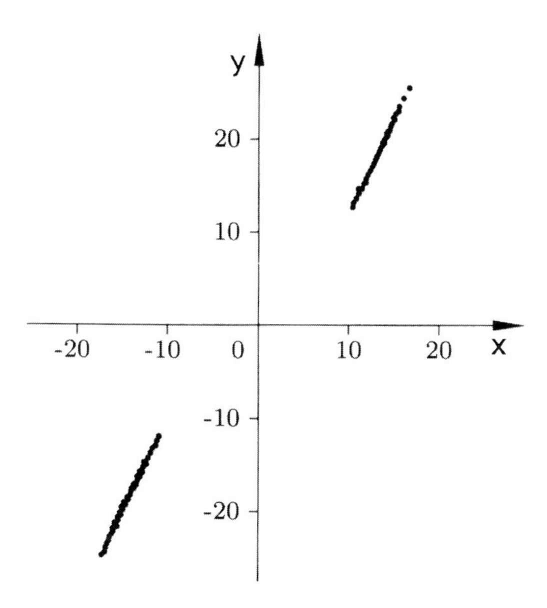

Fig. 7.6 Poincaré section (defined by the plane $z = r - 1$) of the Lorenz attractor (Equations 7.11–7.13; $r = 28$, $\sigma = 10$, $b = 8/3$).

Another example is the driven Brusselator

$$dx/dt = x^2\, y - b\, x - x + a + c\, \cos(\phi)\ , \tag{7.14}$$

$$dy/dt = b\, x - x^2\, y\ , \tag{7.15}$$

$$d\phi/dt = \omega\ , \tag{7.16}$$

which describes chemical oscillations and was one of the numerical prototypes used in Brussels by Ilya Prigogine in part of the work that lead to his Nobel prize. A stroboscopic plot at a fixed phase ϕ of the driving term is equivalent to a Poincaré section at fixed ϕ. Such a Poincaré section yields the four islands shown in Fig. 7.7. The x-values of island 2 yield the map shown in Fig. 7.8, which is well fitted by a parabola [51, 52].

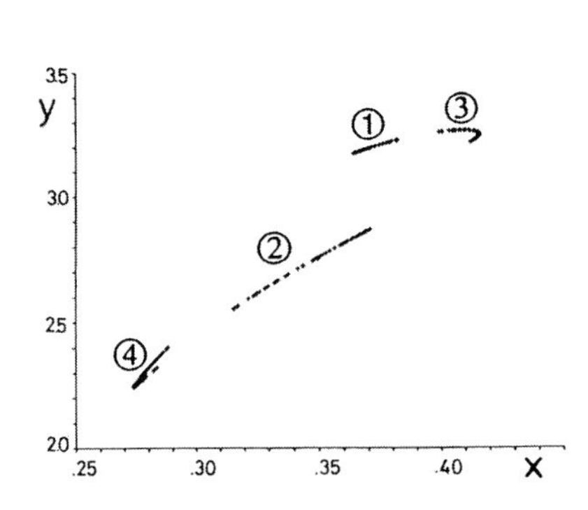

Fig. 7.7 Poincaré section (defined by a fixed ϕ) of the driven Brusselator (Equations 7.14–7.16; $a = 0.4$, $b = 1.2$, $c = 0.05$, $\omega = 0.8$). (Adapted from [51].)

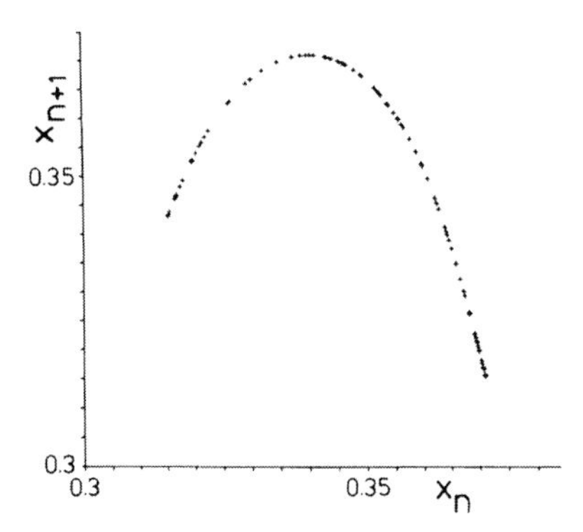

Fig. 7.8 Next-x plot on island 2 of the Poincaré section shown in Fig. 7.7. (Adapted from [51].)

So far, this section has been devoted only to theoretical equations. However, the derivation of discrete maps with Poincaré sections of continuous systems is also possible using experimental data. To address this, let us first consider the so-called phase space reconstruction introduced by the Dutch mathematician Florins Takens [53]: given a measured time series $x(t)$ and a time delay τ, one can construct an attractor by considering $x(t)$, $x(t + \tau)$, $x(t + 2\tau)$,.... $x(t + n\,\tau)$ in an $(n + 1)$-dimensional space. The surprise is that the attractor "reconstructed" in this way has (for appropriate values of τ and sufficiently large n) the same overall properties as the attractor one would have measured if not just $x(t)$ was considered, but other independent variables too. This is extremely important since often only one variable $x(t)$ is accessible to the experimentalist. Moreover,

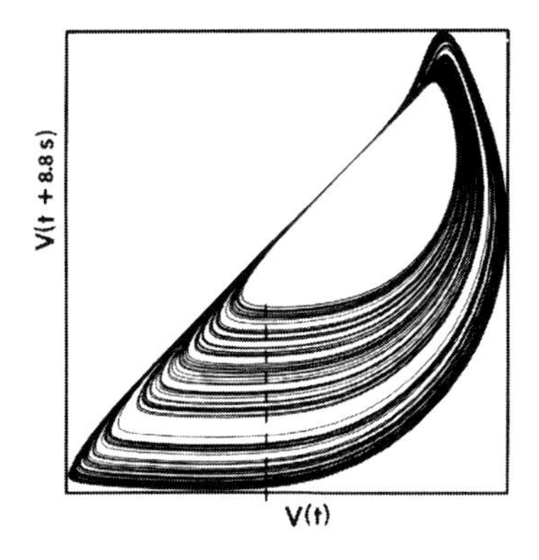

Fig. 7.9 Attractor (projected on a plane) as obtained from experiments with the Belousov–Zhabotinsky reaction [54].

$x(t)$ may have been measured a long time ago, as often happens e.g. in astronomy or ecology; thus there is absolutely no way of getting further independent variables. As an example, Fig. 7.9 shows (for $\tau = 8.8$ seconds), the projection of a 3D-attractor on the $V(t) - V(t+\tau)$-plane, where $V(t)$ is the potential at a bromide sensitive electrode immersed in a chamber where the Belousov-Zhabotinsky reaction took place [54]. (More details on this reaction are given in Section 8.5.)

The attractor in Fig. 7.9 is subject to a Poincaré section, as indicated by the dashed line at a fixed value of $V(t)$. The quasi-1D set resulting from this section is shown in Fig. 7.10. Successive points x_n and x_{n+1} on that quasi-1D-set are plotted in Fig. 7.11. This latter plot is well fitted by the map given in Section 8.5.

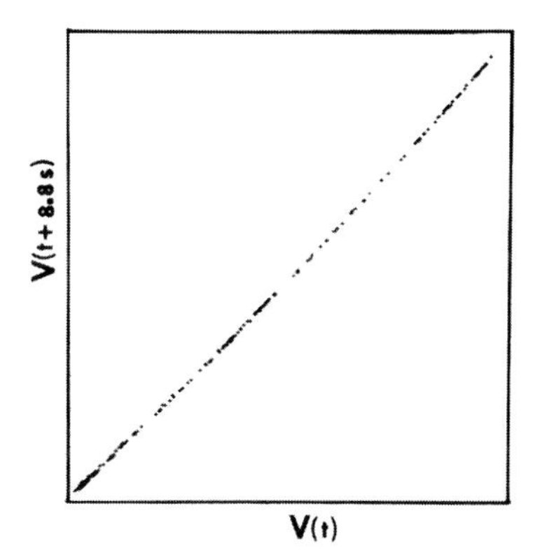

Fig. 7.10 Poincaré section along the dashed line of Fig. 7.9 [54].

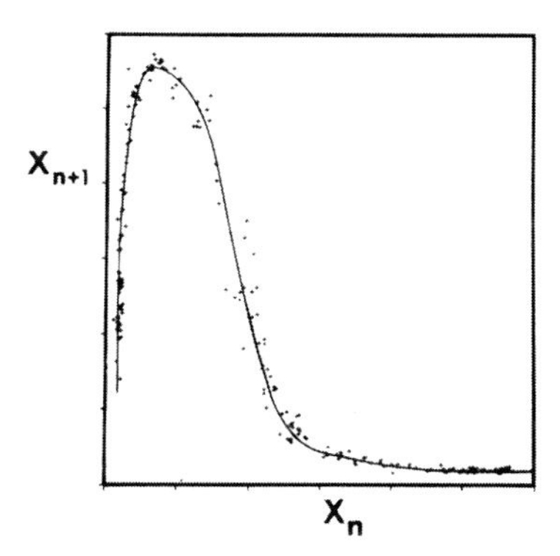

Fig. 7.11 Next-x plot of points on the Poincaré section shown in Fig. 7.10 (points). Straight line: fitted curve corresponding to Equation 8.12 [54].

7.4 Analytical integration

In exceptional cases the differential equations describing a physical system can be integrated analytically, yielding discrete maps. An example is the so-called kicked rotator, which will now be discussed.

Consider a particle rotating on a plane, the angle of rotation being ϕ. The force on the particle consists of periodical "kicks" (period T) described by delta-functions. The equation of motion is

$$\frac{d^2\phi}{dt^2} + \gamma\frac{d\phi}{dt} = K\, f(\phi) \sum_{n=1}^{\infty} \delta(t - nT) \ . \tag{7.17}$$

Dividing the equation by the moment of inertia elliminated the latter. The second term describes the friction. This non-autonomous second-order equation can be written as the following set of three autonomous first-order equations:

$$\frac{dx}{dt} = y \ , \tag{7.18}$$

$$\frac{dy}{dt} = -\gamma y + K\, f(x) \sum_{n=1}^{\infty} \delta(z - nT) \ , \tag{7.19}$$

$$\frac{dz}{dt} = 1 \ , \tag{7.20}$$

where $x = \phi$, $y = \frac{d\phi}{dt}$ and $z = t$. The following discrete variables are introduced: $x_n = x(nT)$, $y_n = y(nT)$. Multiplication of Equation (7.19) with $e^{\gamma t}$ yields

$$\frac{d[e^{\gamma t} y]}{dt} = e^{\gamma t} K f(x) \sum_{n=1}^{\infty} \delta(t - nT) \ . \tag{7.21}$$

Integration of this equation from $nT - \epsilon$ to a time t restricted by $nT < t < (n+1)T - \epsilon$, taking the limit $\epsilon \to 0$ and solving for $y(t)$ yields

$$y(t) = e^{-\gamma(t-nT)}[y_n + K\, f(x_n)] \ . \tag{7.22}$$

Inserting this equation on the right side of Equation (7.18) and proceeding analogously as above renders

$$x(t) = x_n + [1 - e^{-\gamma(t-nT)}]\frac{y_n + K f(x_n)}{\gamma} \ . \tag{7.23}$$

In particular, for $t = (n+1)T - \epsilon$ and taking the limit $\epsilon \to 0$, one obtains the two-dimensional map

$$y_{n+1} = e^{-\gamma T}[y_n + K\, f(x_n)] \ , \tag{7.24}$$

$$x_{n+1} = x_n + (1 - e^{-\gamma T})\frac{y_n + K\, f(x_n)}{\gamma} \ . \tag{7.25}$$

Setting $T = 1$, inserting $y_n + K f(x_n)$ (as obtained by Equation (7.24)) into

Equation (7.25), solving for y_{n+1} and shifting the index n to $n-1$ yields

$$y_n = \frac{\gamma(x_n - x_{n-1})}{e^\gamma - 1}. \tag{7.26}$$

Inserting y_{n+1} and y_n from this equation into Equation (7.24) gives

$$x_{n+1} + e^{-\gamma} x_{n+1} = (1+e^{-\gamma})x_n + (1-e^{-\gamma})K\frac{f(x_n)}{\gamma} . \tag{7.27}$$

If one now makes the following particular choice for $f(x_n)$:

$$(1+e^{-\gamma})x_n + (1-e^{-\gamma})K\frac{f(x_n)}{\gamma} = 1 - ax^2 , \tag{7.28}$$

one obtains

$$x_{n+1} = 1 - ax^2 - e^{-\gamma} x_{n-1} . \tag{7.29}$$

Defining $b = -e^{-\gamma}$ and introducing a new variable $y_n = b\,x_{n-1}$, one obtains the well-known Hénon map [55]

$$x_{n+1} = -a\,x_n^2 + y_n + 1 , \tag{7.30}$$

$$y_{n+1} = b\,x_n . \tag{7.31}$$

In Section 9.7 we will come back to a slightly modified version of this map. The map given by Equations (7.30) and (7.31) is dissipative and yields a chaotic attractor, e.g. for $a = 1.4$ and $b = 0.3$. This attractor is shown in Fig. 7.12a, along with successive magnifications of subsets indicated by squares (Figs. 7.12b, 7.12c and 7.12d). The self-similarity seen in these magnifications is an aspect of the fractal nature of the attractor. The fractal geometry of dissipative chaotic systems has led them to be called "strange attractors".

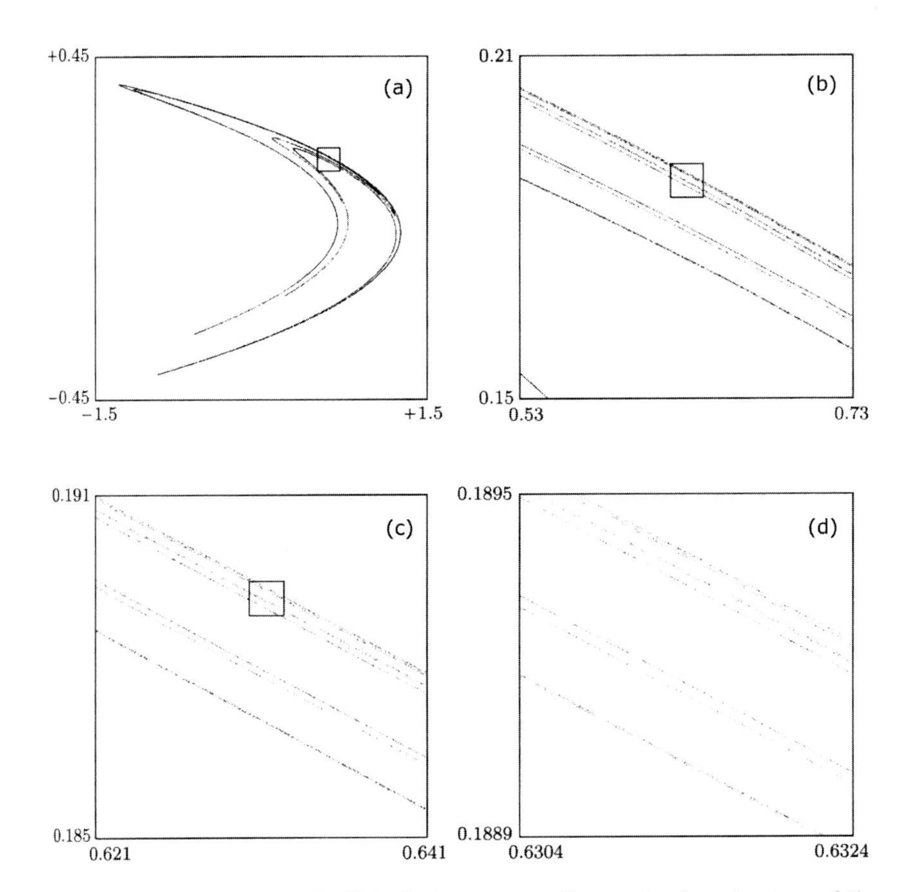

Fig. 7.12 Demonstration of self-similarity corresponding to the fractal nature of the Hénon attractor (a). Successive magnifications of the subsets indicated by squares (b,c and d).

Chapter 8

Maps with Scientific Applications

8.1 A kicked electronic oscillator

In the realm of electronics [56] one encounters oscillators that are driven by periodical impulsive signals ("kicks"). A simple description of such an oscillator is given by

$$\frac{dx}{dt} = s\,x(1 - x^2 - y^2) - y + 2\alpha \sum_{n=-\infty}^{\infty} \delta(t - nT) , \tag{8.1}$$

$$\frac{dy}{dt} = x + s\,y(1 - x^2 - y^2) , \tag{8.2}$$

where x and y are phase variables and the delta-functions describe the kicks. Assuming full relaxation of the oscillator in the time between kicks, the 1D map

$$\theta_{n+1} = \tan^{-1}\left[\frac{\sin(\theta_n + T)}{(2\alpha + \cos(\theta_n + T)}\right] \tag{8.3}$$

is derived [56], where $\theta_n = \tan^{-1}\left(\frac{y_n}{x_n}\right)$. y_n and x_n are the values of the variables inmediately after the nth kick.

Figure 8.1 shows λ on the plane defined by the parameters α (ordinate; kick amplitude) and T (abscissa; kick period). Figure 8.2 shows the effect of alternating T as AABB AABB... holding $\alpha = 0.7$.

Fig. 8.1 Equation (8.3): α versus T. $n_{\text{prev}} = 100$, $n_{\max} = 200$, $x_0 = 0.5$, D-shading. LL:(-8.3, -1.15), UL:(-8.3, 1.15), LR:(2, -1.15)

Fig. 8.2 Equation (8.3): $\alpha = 0.7$. B versus A. T:AABB AABB..., $n_{\text{prev}} = 100$, $n_{\max} = 200$, $x_0 = 0.5$, D-shading. LL:(0, 0), UL:(0, 11.6), LR:(12.2, 0)

8.2 Cardiac oscillations

The heart muscle is normally driven by a periodic electrical input originating in the sinoatrial node. If this stimulation has a pathologically high frequency, not only periodical but also chaotic rhythms result. Chaotic rhythms correspond to the lethal ventricular fibrillation, which was investigated by Michael Guevara in his doctoral thesis at McGill University in Montreal (1984). He showed that the measured dynamics of the heart can be described by the following one-dimensional map:

$$x_{n+1} = A - B_1\, e^{-t_n/\tau_1} - B_2\, e^{-t_n/\tau_2} \,, \tag{8.4}$$

where $t_n = k_n r - x_n$. x_n is the duration of the action potential in the heart cell membrane. r is the period of the stimulation. k_n is the smallest integer such that $k_n r - x_n > t_{\min}$, where $t_{\min}$ is the minimum recovery time.

Figure 8.3 shows λ as a function of $t_{\min}$ (ordinate) and r (abscissa). A periodic alternation of r as ABAB..., holding $t_{\min}$ at 53.5, yields Plate 2.

Fig. 8.3 Equation (8.4): $A = 270$, $B_1 = 2441$, $B_2 = 90.02$, $\tau_1 = 19.6$, $\tau_2 = 200.5$, $t_{\min}$ versus r. $n_{\text{prev}} = 100$, $n_{\max} = 200$, $x_0 = 5$. L-shading. LL:(0, 20), UL:(0, 140), LR:(150, 20)

Fig. 8.4 Equation (8.5): $\alpha = 1$. B versus A. r:AB AB..., $n_{\text{prev}} = 600$, $n_{\max} = 600$, $x_0 = 2$. D-shading. LL:(0, 0), UL:(0, 15), LR:(15, 0)

A simpler approach to the study of heart arrhythmias was given by Stone [57]. In fact, he pointed out that the map

$$x_{n+1} = \sin(\alpha\, x_n) + r \tag{8.5}$$

describes the dynamics of stimulated chicken heart cells, as determined by Guevara *et al.* [58], strikingly well. Holding α constant, Fig. 8.4 is obtained by modulating r as ABAB... Another cardiac model is given in Section 9.6.

8.3 Turbulent diffusion

In a turbulent fluid flow, so-called "super diffusion" can occur, i.e. the mean square displacement (variance) grows faster than linear in time. A map that mimics this behaviour was formulated by Okubo *et al.* [59] as

$$x_{n+1} = x_n + a\,|x_n|^p \sin(x_n)\ . \tag{8.6}$$

Figure 8.5 shows λ for the parameter a alternating as ABAB.... These calculations ($p = 0.3334$) yield a variance growing as $t^{1.5}$.

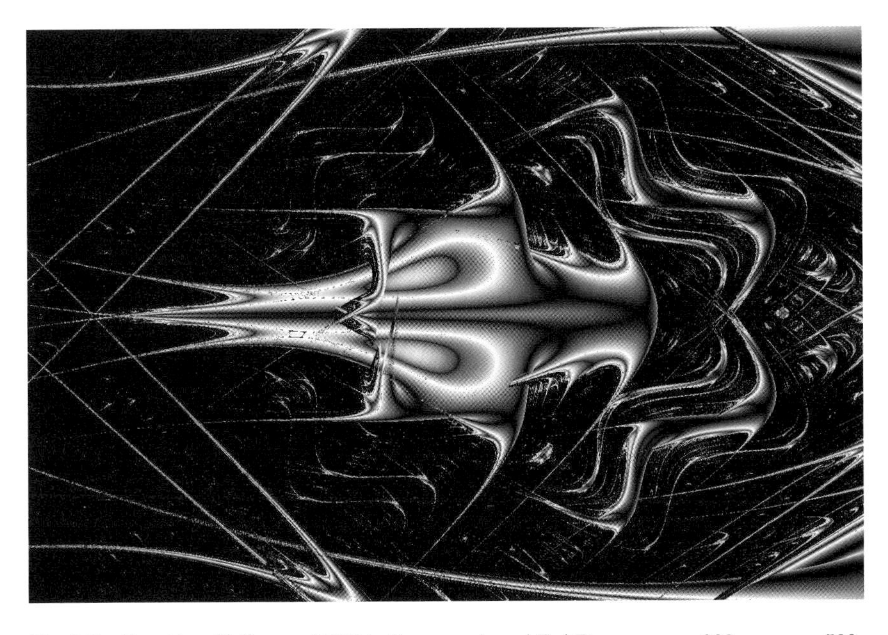

Fig. 8.5 Equation (8.6): $p = 0.3334$. B versus A. a:AB AB..., $n_{\text{prev}} = 200$, $n_{\max} = 500$, $x_0 = 0.5$. D-shading. LL:(2.48, 2.8), UL:(2.8, 2.48), LR:(2, 2.32)

8.4 A traffic model

Traffic often alternates between a "passive" state (extremely slow motion) and an "active" state (normal, fast motion). This alternation has been modelled by Erramilli *et al.* [60] with the so-called double-intermittency map

$$x_{n+1} = \begin{cases} \epsilon_1 + x_n + C_1 x_n^m & \text{for} \quad 0 \le x_n < d \\ -\epsilon_2 + x_n - C_2(1-x_n)^m & \text{for} \quad d \le x_n < 1 \end{cases}, \tag{8.7}$$

where $C_1 = (1-\epsilon_1-d)/d^m$ and $C_2 = -(\epsilon_2 - d)/(1-d)^m$. The statistical distribution of the x_n obtained from this map corresponds to the distributions of traffic activity. Periodic alternations of d yield Figs. 8.6 and 8.7.

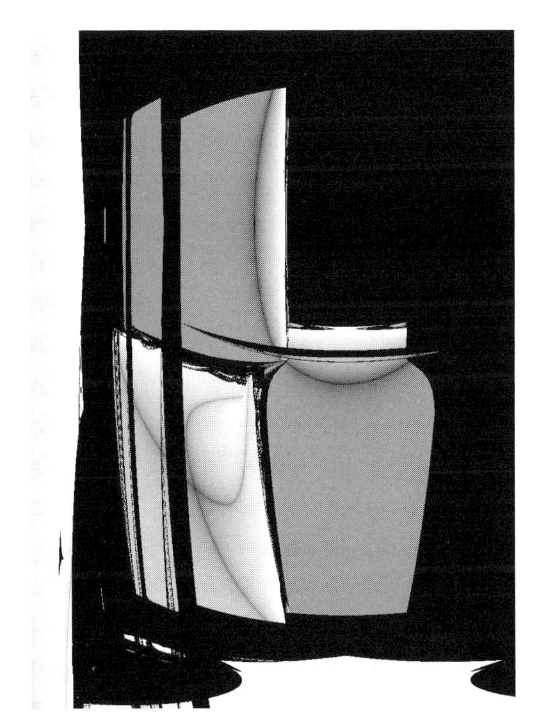

Fig. 8.6 Equation (8.7): $\epsilon_1 = \epsilon_2 = 0.3$, $m = 3.1$. B versus A. d:AB AB..., $n_{\text{prev}} = 100$, $n_{\max} = 200$, $x_0 = 0.5$. L-Shading. LL:(0.845, 0.868), UL:(0.845, 1.1), LR:(1.135, 0.868)

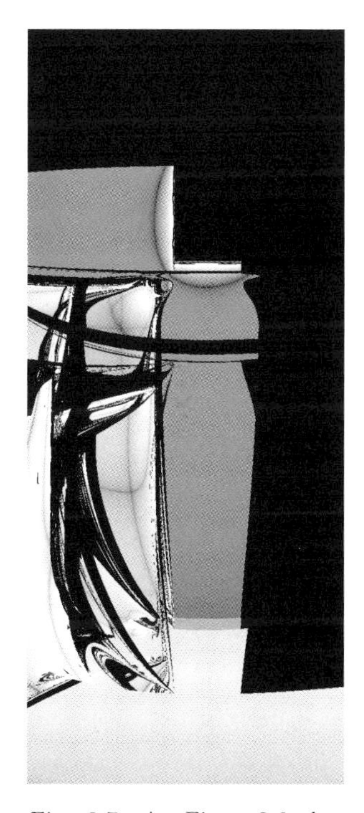

Fig. 8.7 As Fig. 8.6, but d:AA BB AA BB..., $n_{\text{prev}} = 100$, $n_{\max} = 200$, $x_0 = 0.5$. L-shading. LL:(0.8451, 0.56), UL:(0.8451, 1.195), LR:(1.18, 0.56)

8.5 The Belousov-Zhabotinsky reaction

In 1950, the Soviet biophysicist Boris Belousov discovered, purely by chance while doing metabolic experiments, that a chemical reaction can oscillate with constant amplitude. His report was rejected by a scientific journal with the remark that it was clearly impossible for a reaction not to go to thermodynamic equilibrium. After that, Belousov retired from scientific work. In 1961 the investigations were reassumed by Anatol Zhabotinsky, but it was not until a conference in Prague in 1968 that the existence of such oscillations was accepted by the scientific community.

In a simplified way, the oscillations of the Belousov-Zhabotinsky reaction (BZ) can be described as follows. In a first stage of the process, an autocatalytic increase — an "explosion", so to say — of bromous acid $HBrO_2$ occurs: one molecule leads to two, two to four, and so on:

$$HBrO_2 + BrO_3^- + H^+ \rightleftharpoons 2BrO_2 + H_2O \ , \tag{8.8}$$

$$2BrO_2 + Me^{red} + 2H^+ \rightleftharpoons 2Me^{ox} + 2HBrO_2 \ . \tag{8.9}$$

BrO_3^- is the bromate ion. Me^{red} and Me^{ox} are the reduced and the oxydized states of a metal, such as iron, cerium or ruthenium. This metal is incorporated in a complex that acts both as catalyst and as coulour indicator. The "explosion" is stopped by the consumption of Me^{ox} and by the production of bromide ions Br^-, which react with $HBrO_2$:

$$Me^{ox} + MA + BrMA \rightleftharpoons Me^{red} + Br^- + \text{org. products} \ , \tag{8.10}$$

$$Br^- + HBrO_2 + H^+ \rightleftharpoons 2HBrO \ . \tag{8.11}$$

Here, MA is malonic acid $CH_2(COOH)_2$. As a result of the two reactions (8.10) and (8.11), $HBrO_2$ and its inhibitor Br^- are largely consumed, while Me^{red} becomes available again. Then the "explosion", given by the reactions (8.8) and (8.9), restarts. The different colours of Me^{red} and Me^{ox} show how the reaction goes back and forth. Such an oscillatory reaction is referred to as a "chemical clock". Biorhythms (see Section 4.1) are chemical clocks ocurring in metabolic processes, the catalysts being enzymes (see [19]).

The experiments with the BZ reaction by Harry Swinney's group in Austin, Texas [61] were done in a stirred chamber into which fresh chemicals were continuously injected, while the solution (with reactants and products) left the chamber at the same rate. To their surprise they found that if they changed the "residence time" τ (τ =chamber volume/flux rate), then the oscillations became chaotic, i.e. the chemical clock "ticked" with unpredictable periods and amplitudes. Later on, in 1983 [54], scientists from the same group found that the chaotic oscillations can be described by a 1D plot, as described in Section 7.3. This plot [62, 63] is described by the map

$$x_{n+1} = \begin{cases} [(\frac{1}{8} - x)^{\frac{1}{3}} + r]e^{-x} + b & \text{for} \quad 0 \leq x < \frac{1}{8} \\ [(x - \frac{1}{8})^{\frac{1}{3}} + r]e^{-x} + b & \text{for} \quad \frac{1}{8} \leq x < \frac{3}{10} \\ c[10x\, e^{-10x/3}]^{19} + b & \text{for} \quad \frac{3}{10} \leq x < 1 \end{cases} , \tag{8.12}$$

where $b = 0.0232885279$ and $c = 0.063633 + 0.1137635r$. Alternating r as ABAB... leads to the diagram on Plate 3.

8.6 The international arms race

In 1984, the journal *Nature* featured a surprising article by Alvin Saperstein containing a model for the outbreak of war [64]. It was assumed there that a bilateral arms race between two competing countries occurs in stages. These stages, which were designated by an integer n, may indicate successive years, successive budget cycles and the like. The fraction of its resources that each country devoted to armament in stage n was called x_n and y_n. x_{n+1} was assumed proportional to y_n and y_{n+1} proportional to x_n. The reason is that a nation's armament in the next stage depends upon its perception of its opponent's armaments at the present stage. In addition, the perceived threat depends upon how much of the opponent's resources remain potentially convertible to armaments in the future. Indeed, if the opponent country devotes all of its resources to arms, there is no need for concern about a future increase in the threat it poses. Hence, x_{n+1} was assumed proportional to $1 - y_n$, and conversely. Thus,

$$x_{n+1} = r\, y_n(1 - y_n) , \tag{8.13}$$

$$y_{n+1} = b\, x_n(1 - x_n) . \tag{8.14}$$

$\lambda > 0$ occurs above a curve calculated by Saperstein on the r-b-plane. Saperstein then proceeded to look at the ratio of the national military expenditure to the gross national product in France, Germany, Italy, Great Britain and the Soviet Union for 1935, 1936 and 1937. With these ratios he could estimate r and b for all pairs of countries. For opposing countries, he found r and b within or very near the region with $\lambda > 0$, giving an explanation for the outbreak of the Second World War. In contrast, he calculated r and b for the Soviet-USA arms race in 1984, obtaining values well within the region with $\lambda < 0$ (stability).

Figure 8.8 shows λ for r fluctuating as BABA... at fixed b. One observes regions with $\lambda > 0$ (black) and regions with $\lambda < 0$ (lighter) that are small and very close to one another. In other words, one sees that fluctuating conditions can cause extreme difficulties for the assertion that the situation is stable ($\lambda < 0$).

Fig. 8.8 Equations (8.13, 8.14): $b = 3.865$. B versus A. r:BA BA..., $n_{\text{prev}} = 100$, $n_{\max} = 200$, $x_0 = y_0 = 0.4$. D-shading. LL:(0, 0), UL:(0, 4), LR:(2.95, 0)

8.7 The predictor-corrector equation for the kicked rotator

The kicked rotator was presented in Section. 7.4 as an example of a continuous physical system that can be described by a discrete map via analytical integration.

Assuming, in Equations (7.24) and (7.25), $T = 1$, as well as a force in a constant direction, i.e. $f(x_n) = \sin(x_n)$, and considering the limit of zero friction ($\gamma \to 0$), we obtain the standard or Chirikov map:

$$y_{n+1} = y_n + K \sin(\phi_n) , \tag{8.15}$$

$$\phi_{n+1} = \phi_n + y_{n+1} , \tag{8.16}$$

where $y_n = \frac{d\phi_n}{dt}$. Miller [65] applied a predictor-corrector difference scheme to refine this description of the kicked rotator. As a result, he modified the standard map to

$$y_{n+1} = y_n + \frac{K}{2} [\sin(\phi_n) + \sin(\phi_n + y_n)] , \tag{8.17}$$

$$\phi_{n+1} = \phi_n + \frac{(y_n + y_{n+1})}{2} . \tag{8.18}$$

If now the force K is modulated periodically as BABA..., one obtains Fig. 8.9.

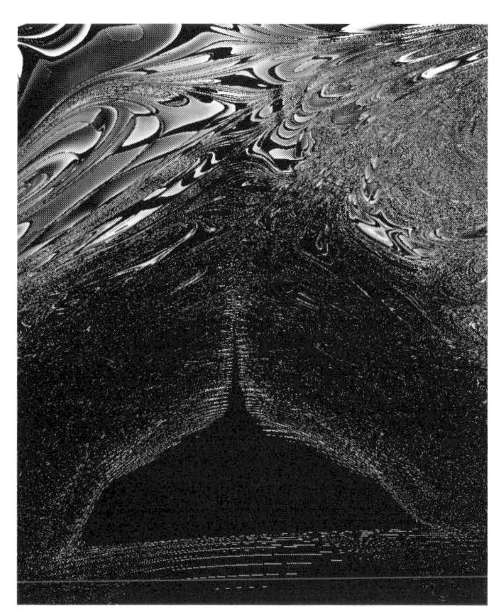

Fig. 8.9 Equations (8.17, 8.18): B versus A. K:BA BA..., $n_{\text{prev}} = 100$, $n_{\text{max}} = 200$, $x_0 = y_0 = 0.4$. D-shading. LL:(-0.5, 0.215), UL:(0.412, 1.1), LR:(0.248, -0.556)

8.8 Interdependent economies

Shahriar Yousefi *et al.* [66] considered the markets for two goods that satisfy the "perfect substitutability assumption". This assumption implies that goods of a given type supplied by one country are perfect substitutes for the same sort of goods supplied by another country. They derived the following simple equations:

$$x_{n+1} = \mu_1 \, x_n(1 - x_n) + \gamma_1 \, y_n \ , \tag{8.19}$$

$$y_{n+1} = \mu_2 \, y_n(1 - y_n) + \gamma_2 \, x_n \ , \tag{8.20}$$

where x_n and y_n are proportional to the market clearing prices of the goods. μ_1 and μ_2 are complicated functions of the unit costs, interest rates, taxes and other parameters, which are assumed to be fixed. γ_1 and γ_2 are the trade policy parameters. Here, the symmetrical case is considered, i.e. $\gamma_1 = \gamma_2 = \gamma, \quad \mu_1 = \mu_2 = \mu$. Obviously, if there was no trade between the economies, $\gamma = 0$. Assuming a constant γ and allowing μ to alternate as BABA..., one gets Fig. 8.10.

Fig. 8.10 Equations (8.19, 8.20): $\gamma_1 = \gamma_2 = 0.43$. B versus A. $\mu_1 = \mu_2$:BA BA..., $n_{\text{prev}} = 100$, $n_{\text{max}} = 300$, $x_0 = y_0 = 0.4$. D-shading. LL:(2, 2.40625), UL:(2.24375, 2.65), LR:(2.40625, 2)

8.9 Asymmetric economic competition

Feichtinger's microeconomic model (see [67, 68]) describes the interaction of two firms, X and Y, with yearly sales x_n and y_n, which compete in the same market for goods. The firms invest according to different strategies: X invests if it has an advantage over Y, while Y invests if it is in a disadvantageous position with respect to X. In other words, both firms invest if $x_n < y_n$. The model reads

$$x_{n+1} = (1-\alpha)x_n + a/[1+e^{-c(x_n-y_n)}] , \quad (8.21)$$

$$y_{n+1} = (1-\beta)y_n + b/[1+e^{-c(x_n-y_n)}] . \quad (8.22)$$

The constants $1-\alpha$ and $1-\beta$ are the rates of the decay in sales in both firms in the absence of investments. c (the so-called "elasticity" of the strategies) is a large number (around a thousand times larger than x_n and y_n) that acts as a switch. In fact, for $x_n < y_n$, the investments will be practically equal to zero, while for $x_n > y_n$, the investment of X (or Y) will be nearly equal to a (or b).

It was demonstrated that the necessary conditions for chaos to occur are $\alpha < \beta$ and $a < b$. Figure 8.11 shows λ for a periodical change of a, i.e. of the investment of firm X.

Fig. 8.11 Equations (8.21, 8.22): $\alpha = 0.46$, $\beta = 0.7$, $c = 105$, $b = 4$. B versus A. a:AABAB AABAB..., $x_{\text{prev}} = 500$, $x_{\text{max}} = 1000$, $x_0 = y_0 = 0.6$. D-shading. LL:(0, 0), UL:(0, 3.6), LR:(3.6, 0)

8.10 Economic-ecological feedback in the fishing business

Alan Berryman from Washington State University investigated the interactions between crab fishery investments and crab population [69]. Specifically, he analysed the aperiodic records of the Dungeness crab (*Cancer magister*; it measures up to 25 cm) in northern California between 1950 and 1978.

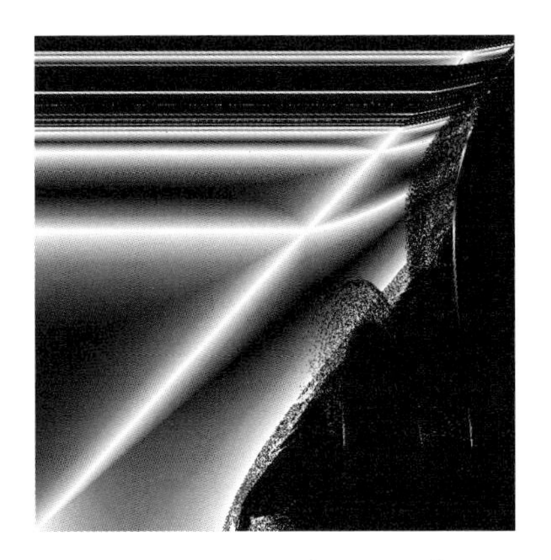

Fig. 8.12 Equations (8.23, 8.24): $b_D = -0.005$, $c_D = -0.05$, $b_P = -0.04$, $c_P = 0.007$. a_P versus a_D. $n_{\text{prev}} = 500$, $n_{\max} = 1000$, $x_0 = y_0 = 0.5$. D-shading. LL:(0, 0), UL:(0, 3.3), LR:(4.4, 0)

Fig. 8.13 Equations (8.23, 8.24): $a_D = 1$, $b_D = -0.005$, $a_P = 0.5$, $b_P = -0.04$. c_P versus c_D. $n_{\text{prev}} = 5$, $n_{\max} = 20$, $x_0 = y_0 = 0.5$. D-shading. LL:(−0.033, 0), UL:(−0.033, 0.105), LR:(0, 0)

The following interactions take place: (i) Crab intra-specific competition reduces reproduction and increases cannibalism; (ii) Crab harvest success in one year increases investments in crab pots for fishing in the next year; and (iii) If the number of pots becomes too high, some are taken away to be used elsewhere. Berrymans's model reads:

$$D_{n+1} = D_n \, e^{a_D + b_D \, D_n + c_D \, P_n} \; , \tag{8.23}$$

$$P_{n+1} = P_n \, e^{a_P + b_P \, P_n + c_P \, D_n} \; . \tag{8.24}$$

D_n is the crab biomass and P_n the number of fishing pots. $b_D < 0$ and $b_P < 0$ correspond to the self-inhibitory effects of D_n and P_n. $c_D < 0$ corresponds to the crab harvesting by pots. $c_P > 0$ represents the investment in pots after harvesting success. $a_P > 0$ describes the investment growth, independently of success. $a_D > 0$ stands for the natural crab reproduction rate.

Figure 8.12 shows λ on the a_D-a_P-plane. Figure 8.13 shows λ for transients on the c_D-c_P-plane. Note that these two figures indicate the influence of fishing policies on the predictability of the ecosystem.

8.11 Laser pulse in a ring cavity

In this section, the evolution of a laser pulse in a ring cavity, as illustrated in Fig. 8.14, is considered [70]. A light pulse enters the partially transmitting

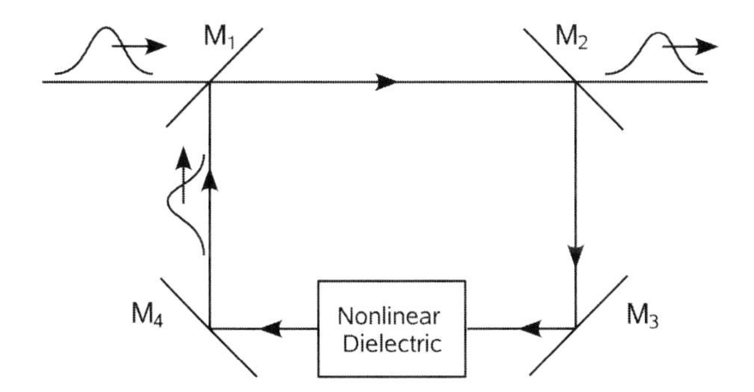

Fig. 8.14 Scheme for the trajectories of a laser pulse in a ring cavity.

mirror M_1. The amplitude and phase of the pulse just to the right of M_1 is defined by z_n, where n indicates the number of round trips of the pulse in the system. Kensuke Ikeda [70] derived the map

$$z_{n+1} = p + B\, z_n \, \exp\left[i\, K - i \frac{\alpha}{1 + |z_n|^2} \right] , \tag{8.25}$$

which is two-dimensional, since $z_n = x_n + i\, y_n$ is a complex number. B is the fraction of energy absorbed by the mirrors M_1 and M_2. M_3 and M_4 are assumed to have 100% reflectivity. K is the round-trip phase shift that would be experienced by the pulse in the absence of the non-linear medium. $-\frac{\alpha}{1+|z_n|^2}$ is the phase shift due to the presence of the non-linear medium.

Setting $p = 1.0$ and $K = 0.4$ (as in [71]), one obtains λ on the plane defined by B (abscissa) and α (ordinate) as shown in Fig. 8.15.

A simpler model for a ring cavity with gain has been proposed by Lu and Tan [72]. It describes a system with three mirrors, the laser frequency

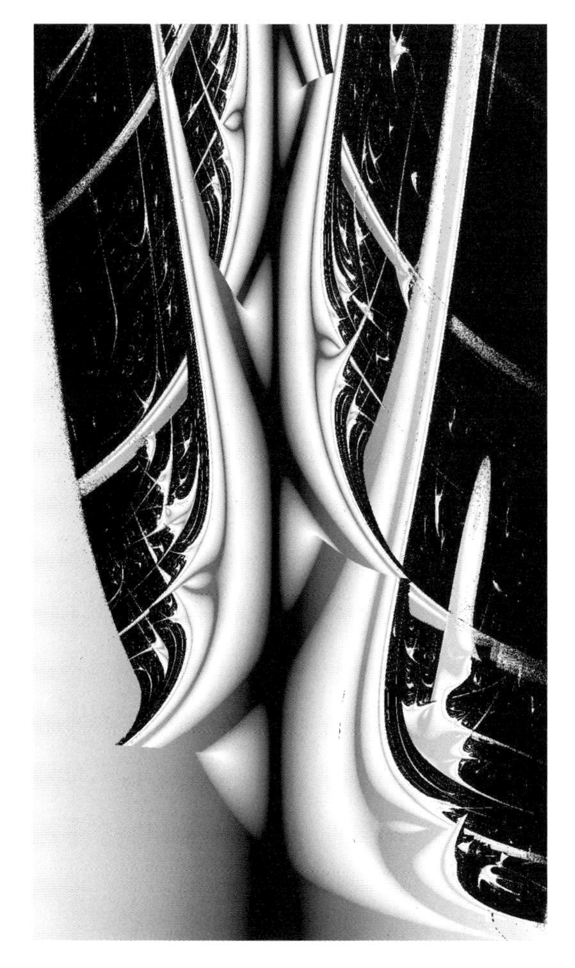

Fig. 8.15 Equation (8.25): $p = 1$, $K = 0.4$. α versus B. $n_{\mathrm{prev}} = 200$, $n_{\mathrm{max}} = 300$, $x_0 = -0.5$, $y_0 = 0$. L-shading. LL:$(-0.8, 4.7)$, UL:$(-0.8, 31)$, LR:$(0.9, 4.7)$

being resonant with the gain medium frequency. The light intensity x_n is described by the map

$$x_{n+1} = x_n \, e^{r/(1+x_n)-b} , \tag{8.26}$$

where r is the gain coefficient of the medium and b stands for the losses in the cavity. Figure 8.16 shows λ for this latter model, as obtained from periodical modulation of r.

Fig. 8.16 Equation (8.26): $b = 11$. B versus A. r:AABB AABB..., $n_{\text{prev}} = 200$, $n_{\max} = 300$, $x_0 = 0.31$. D-shading. LL:(23.5, 32.7), UL:(33.1, 42.5), LR:(32.8, 23.5898)

8.12 Periodically driven elements of antiferromagnetic lattices

Angelini [73] simulated antiferromagnetism in a lattice considering the following map in each lattice element:

$$x_{n+1} = \begin{cases} -\frac{r}{3}\exp\left[b\left(x_n + \frac{1}{3}\right)\right] & \text{if} \quad x_n < -\frac{1}{3} \\ r\,x_n & \text{if} \quad -\frac{1}{3} < x_n < \frac{1}{3} \\ \frac{r}{3}\exp\left[b\left(\frac{1}{3} - x_n\right)\right] & \text{if} \quad x > \frac{1}{3} \end{cases} . \tag{8.27}$$

The maps in each element of a two-dimensional lattice were coupled via

$$x_{n+1}^{i} = F(x_n^i) - \frac{1}{N} \sum_j (F(x_n^j) - x_n^j) \; , \tag{8.28}$$

where

$$F(x_n^i) = (1 - N\, g) f(x_n^i) + g \sum_j f(x_n^j) \; . \tag{8.29}$$

g is the coupling strength and the summations are carried out over a neighbourhood with N elements [74]. The upper index i indicates the site of the element. Magnetic phenomena are described by setting the spin equal to the sign of x_n. Considering the case of decoupled elements and modulating r periodically leads to λ, as shown in Fig. 8.17. The consideration of finite coupling is left until later.

Fig. 8.17 Equation (8.27): $b = 1$. B versus A. r:B^{5}AA B^{5}AA..., $n_{\mathrm{prev}} = 100$, $n_{\mathrm{max}} = 200$, $x_0 = 1$. D-shading. LL:(−15, −4.3), UL:(−15, 11.), LR: (35, −4.3)

8.13 Periodically driven p-n junctions

Carson Jeffries [75] investigated the transition between periodicity and chaos in the following electronic system. A silicon crystal containing fixed donor ions and electrons on one side and fixed acceptor ions and holes on the other side (p-n junction) is connected in series with an inductance, a resistance and a sinusoidal voltage generator.

Poincaré sections of the resulting trajectories in phase space can be well fitted by the two-dimensional map

$$x_{n+1} = [x_n - S(x_n) + 1] - S(r(1 - S(x_n)^2)) - y_n \ , \tag{8.30}$$

$$y_{n+1} = b\, x_n \ . \tag{8.31}$$

$$S(x) = \frac{1}{2}(x + \sqrt{x^2 + b}\,) \ . \tag{8.32}$$

Figure 8.18 shows transient λ's obtained by periodically modulating r.

Fig. 8.18 Equations (8.30–8.32): $b = 0.9$. B versus A r:AAB AAB..., $n_{\rm prev} = 50$, $n_{\max} = 200$, $x_0 = 2.$, $y_0 = 0.4$. D-shading. LL:(0, 0), UL:(0, 5), LR:(2.7, 0)

8.14 Optical bistability in liquid crystals

Hong-jun Zhang *et al.* [76–79] considered a twisted nematic liquid crystal (LC) with a feeding loop consisting of an acoustic delay line or a microprocessor. When the orientation of the molecules at the upper and lower surface of the cell is at $-45°$ or $+45°$ with respect to the crossed polarizers, the transmittance of the LC cell can be approximated by

$$T(V) = H \sin^2 \frac{\pi V}{2V_H} , \tag{8.33}$$

where V is the total voltage applied to the cell. The maximum transmittance H occurs at $V = V_H$. (Note that this approximation is only valid for $|V| < 2V_H$.) The output of light intensity I_2 is related to the input density I_1 by

$$I_2(t) = I_1 \, T(V(H)) . \tag{8.34}$$

$V(t)$ is the sum of the feedback voltage $V_1(t)$ and a constant (adjustable) bias V_B. Given a time delay t_R for the feedback,

$$V_1 = k \, I_2(t - t_R) , \tag{8.35}$$

where k corresponds to the gain of the photodetector and the amplifier in the feedback circuit, the previous equations yield

$$I_2(t) = H I_1 \sin^2 \left(\pi \, k \, I_2 \frac{(t - t_R)}{2V_H} + \frac{\pi \, V_B}{2V_H} \right) . \tag{8.36}$$

Denoting $x(t) = \frac{\pi \, k \, I_2(t)}{2V_H}$ and measuring t in units of t_R , i.e. letting $t = (n+1)t_R$ and $x_n = x(n \, t_R)$, we obtain the approximation

$$x_{n+1} = b \sin^2(x_n + r) , \tag{8.37}$$

where $b = k \, H \, I_1 / 2V_H$ and $r = \pi \, V_B / 2V_H$. Note that the adjustable parameters in the experimental setup are the input density I_1, which is related to b, and the constant bias V_B, which is related to r.

Although the map just derived refers to an application in science, it will be treated in the next chapter on generic systems (Section 9.20) because the performed approximation of the physical features is extremely rough, while the generic properties have been found to be important.

8.15 Host-parasitoid models

Fitting data for insects and their parasitoids led Beddington *et al.* [80] to the following map

$$H_{n+1} = H_n \, \exp \left[r \left(1 - \frac{H_n}{K} \right) - \alpha \, P_n \right] , \tag{8.38}$$

$$P_{n+1} = H_n [1 - e^{-\alpha \, P_n}] . \tag{8.39}$$

H_n and P_n are the population densities of the host and the parasitoid in generation n. r is the reproductive rate and K the carrying capacity of the host. The parameter α is the search efficiency of the parasitoid, which increases its own population and decreases that of the host. Figure 8.19 shows the dependence of λ on r (abscissa) and α (ordinate). Note the strong structural instability of this system in some regions of the parameter space.

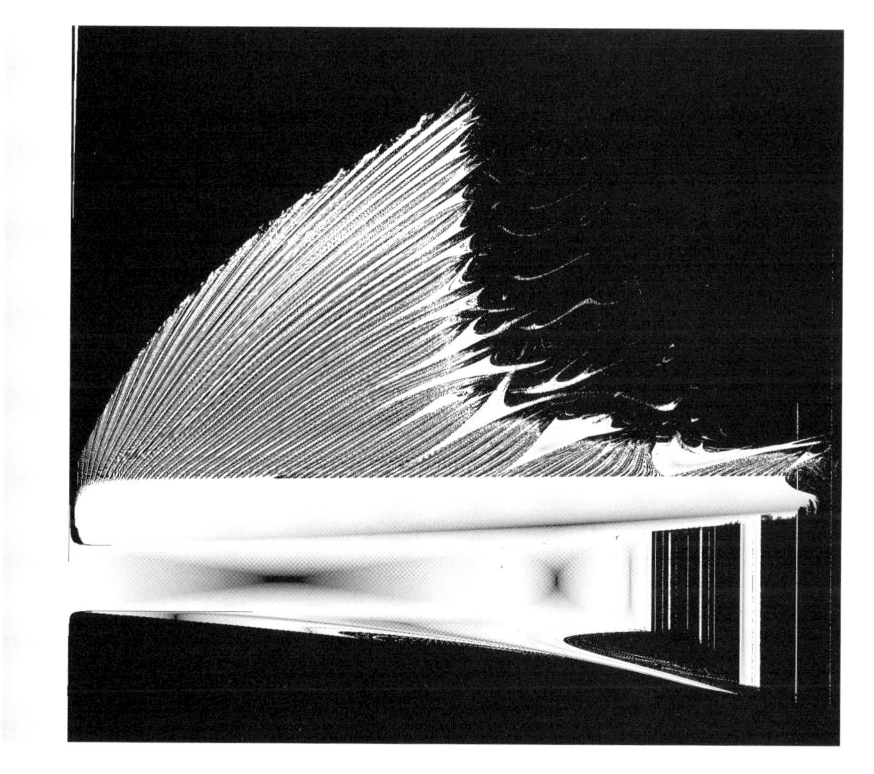

Fig. 8.19 Equations (8.38, 8.39): $K = 2.1$ α versus r. $n_{\text{prev}} = 200$, $n_{\text{max}} = 800$, $x_0 = y_0 = 0.5$. L-shading. LL:(−0.1, −2.4), UL:(−0.1, 8.1), LR:(3.57, −2.4)

Let us now consider a single population x_n. Some data fit better an experimental model

$$x_{n+1} = x_n r e^{-\beta x_n} , \tag{8.40}$$

(see Section 7.2), while other data better fit the logistic model

$$x_{n+1} = r x_n (1 - x_n) \tag{8.41}$$

(see Section 9.2). Replacing the exponential dependence on H_n in Equation

8.38, which would correspond to a model described by Equation 8.40 for the host, by $1 - H_n$, which would correspond to the logistic model, one gets the equations

$$H_{n+1} = \mu H_n (1 - H_n) e^{-\alpha P_n} , \tag{8.42}$$

$$P_{n+1} = P_n [1 - e^{-\alpha P_n}] , \tag{8.43}$$

where $\mu = e^r$. These equations were derived by Solé *et al.* [81] and describe insect data by Hassell and May [82]. They generate Fig. 8.20 when representing λ in the plane defined by α (abscissa) and μ (ordinate).

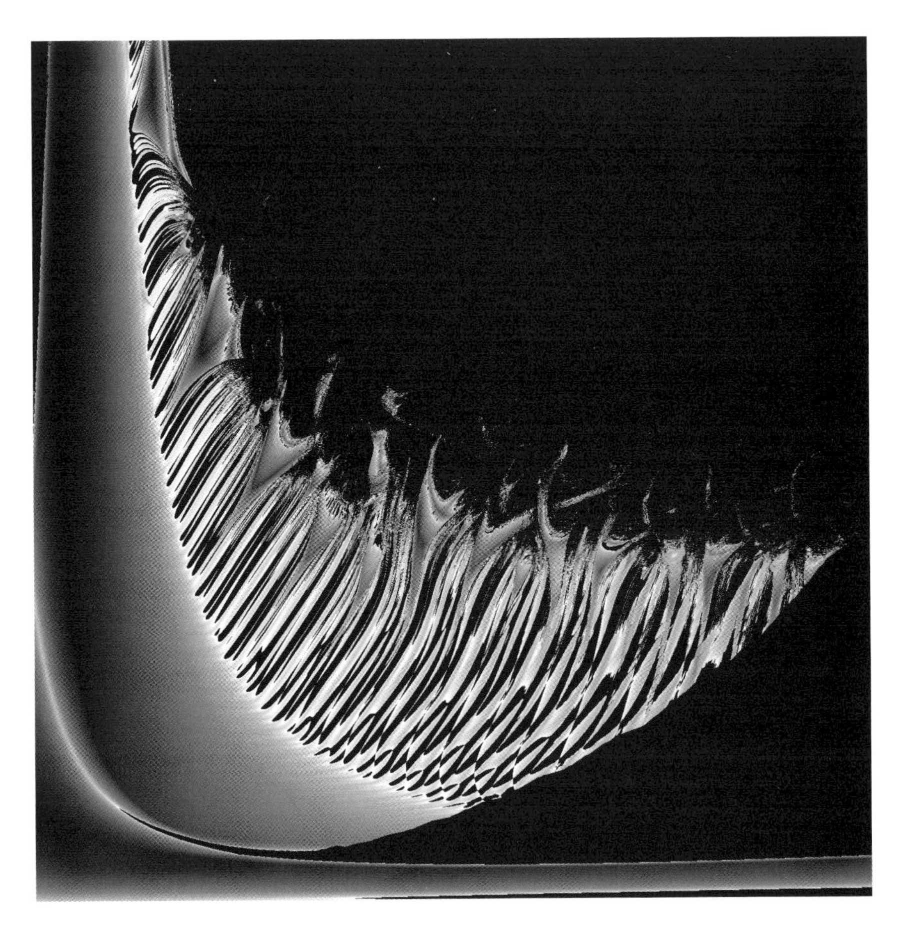

Fig. 8.20 Equations (8.42, 8.43): μ versus α. $n_{\text{prev}} = 100$, $n_{\text{max}} = 200$, $x_0 = 0.5$, $y_0 = 0.3$. D-shading. LL:(1.6, 1), UL:(1.6, 5.8), LR:(15, 1)

8.16 Other predator-prey models

As the logistic equation (see Section 9.2), being — in spite of its simplicity — a generous source of non-linear phenomena, there exists a predator-prey system that is similarly fruitful:

$$x_{n+1} = a\,x_n(1 - x_n - y_n)\ , \tag{8.44}$$

$$y_{n+1} = b\,x_n\,y_n\ . \tag{8.45}$$

x_n and y_n are the fractions of the prey and the predator, respectively. In what follows, some features of these equations are described. Let us set a equal to b. At $a = 1$, a transcritical bifurcation occurs. In fact, the fixed points $(0,0)$ and $(1 - \frac{1}{a}, 0)$ exchange their stability properties. At $a = 2$ a new transcritical bifurcation takes place: $(1 - \frac{1}{a}, 0)$ exchanges stability with the fixed point $(\frac{1}{a}, 1 - \frac{2}{a})$. At the latter fixed point, a Hopf bifurcation occurs for $a = 3$; this leads to oscillations between six points that form a symmetrical polygon. A second Hopf bifurcation, occurring at $a = (11 + \sqrt{85})/6 \approx 3.37$, yields a stable 18-cycle. If one continues to

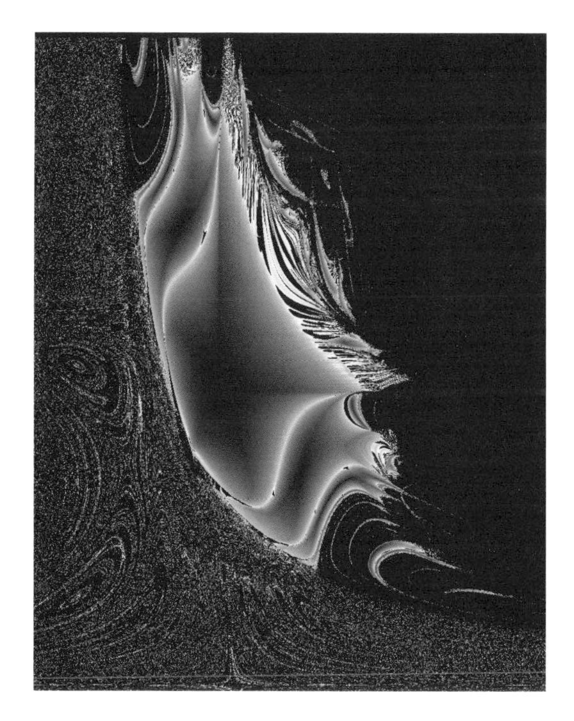

Fig. 8.21 Equations (8.44, 8.45): $b = 3.569985$. B versus A. a:BA BA..., $n_{\text{prev}} = 100$, $n_{\max} = 200$, $x_0 = y_0 = 0.4$. D-shading. LL:$(-0.6, -0.04)$, UL:$(-0.6, 4.5)$, LR:$(6.6, -0.04)$

increase a, period doublings occur (periods 36, 72,...) until chaos appears. At a around 3.43, the chaotic attractor displays six islands around the six points mentioned above.

Setting b constant and alternating a as BABA... one obtains λ, as shown in Fig. 8.21. A slightly more complicated, but symmetrical, predator-prey model is given by

$$x_{n+1} = r\,x_n(1 - x_n) + c\,x_n\,y_n\ , \tag{8.46}$$

$$y_{n+1} = b\,y_n(1 - y_n) - d\,y_n\,x_n\ . \tag{8.47}$$

(See [83].) Setting $c = d$ and constant, one obtains Fig. 8.22, which shows λ on the plane defined by r (abscissa) and b (ordinate).

Fig. 8.22 Equations (8.46, 8.47): $c = d = 0.9$. b versus r. $n_{\text{prev}} = 100$, $n_{\max} = 300$, $x_0 = y_0 = 0.4$. D-shading. LL:(1.75, 3.12), UL:(1.75, 4.3), LR:(3.6, 3.12)

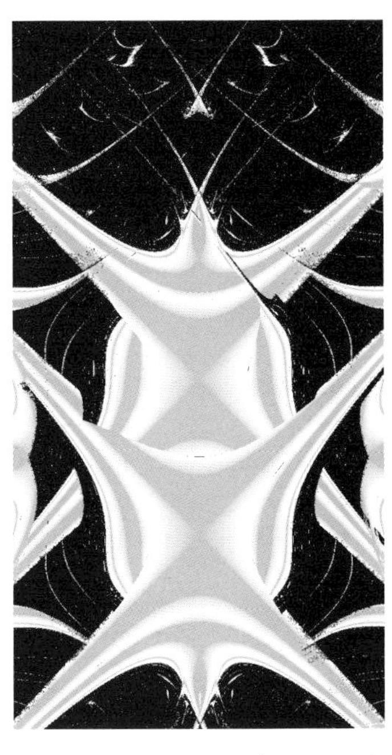

Fig. 8.23 Equations (7.24, 7.25): $T = 1$, $\gamma = 1.5$. B versus A, K:BA BA..., $n_{\text{prev}} = 100$, $n_{\max} = 300$, $x_0 = y_0 = 0.4$. L-shading. LL:(-14, -2.45), UL:(7.45, 19), LR:(-2.45, -14)

8.17 The kicked rotator

In Section 7.4 a two-dimensional map was derived for a rotating particle subject to torque-kicks. These kicks consist of periodic delta-functions (period T), the magnitude of which depend via $f(x_n)$ on the angle x_n. Assuming a force with a constant direction, i.e. setting $f(x_n) = \sin(x_n)$ in Equations (7.24) and (7.25), one obtains Fig. 8.23 for $T = 1$, $\gamma = 1.5$ and K alternating as BABA....

The kicked rotator can be slightly modified by adding a constant torque to the delta-kicks. In the formulation by Zaslavsky ([84], see also [85]), the resulting map then reads

$$x_{n+1} = [x_n + r(1 + \mu\, y_n) + \epsilon\, r\, \mu\, \cos(2\pi\, x_n)] \bmod 1\ , \tag{8.48}$$

$$y_{n+1} = e^{-\gamma}[y_n + \epsilon\, \cos(2\pi\, x_n)]\quad , \tag{8.49}$$

where $\mu = (1 - e^{-\gamma})/\gamma$. Setting constant values for ϵ and γ and modulating r as BABA... one obtains Figs. 8.24 and 8.25.

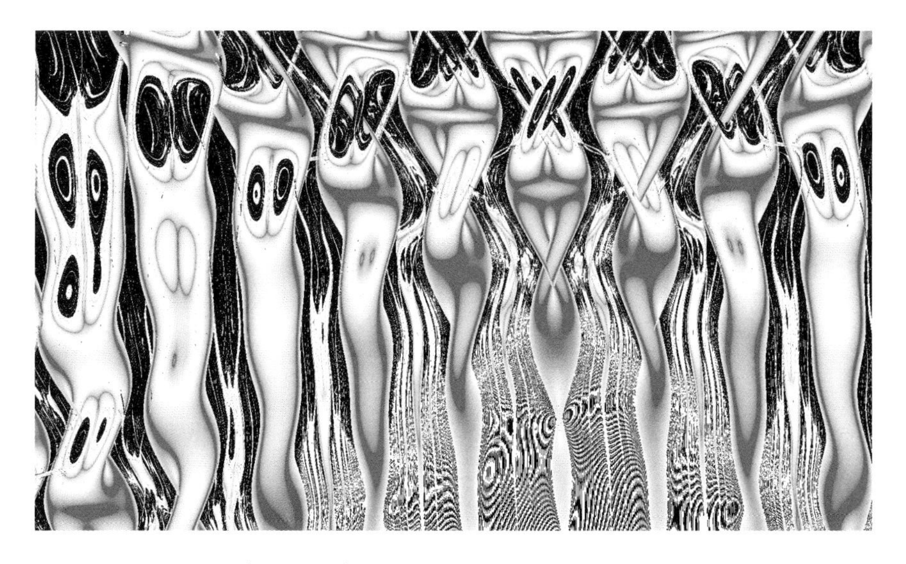

Fig. 8.24 Equations (8.48, 8.49): $\epsilon = 0.3$, $\gamma = 3$. B versus A. r:BA BA..., $n_{\text{prev}} = 100$, $n_{\max} = 200$, $x_0 = y_0 = 0.4$. L-shading. LL:(−3.215, −2.45), UL:(−6.35, 0.9325), LR:(1.4325, 1.85744)

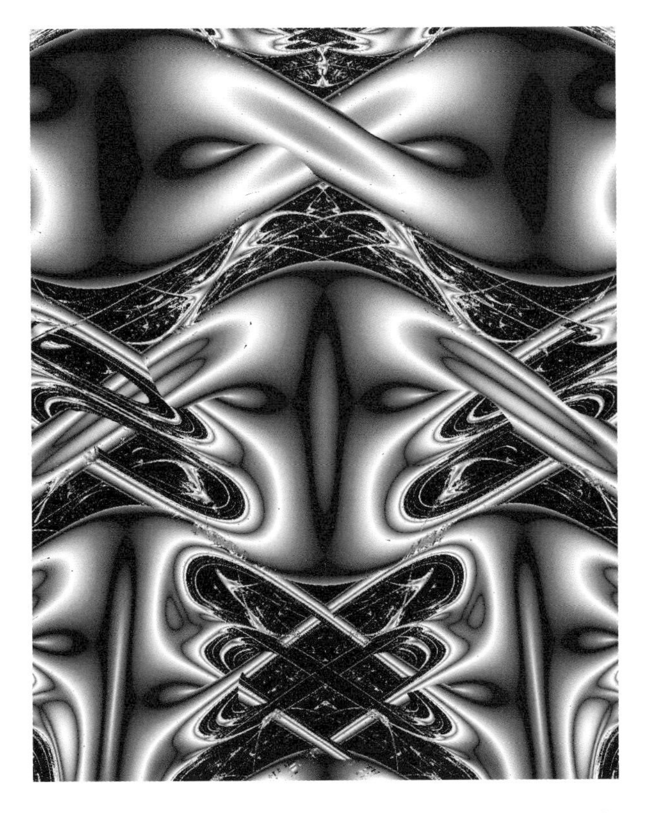

Fig. 8.25 Equations (8.48, 8.49): As Fig. 8.24, but on a different subset of the A-B-plane and $n_{\text{prev}} = 200$, $n_{\max} = 400$. D-shading. LL:(−4.4, −2.995), UL:(−2.995, −1.59), LR:(−2.995, −4.4)

8.18 The buckled beam, Pohl's wheel and Holmes' map

There are two devices which have become popular in classroom demonstrations of deterministic chaos: the buckled beam and Pohl's wheel.

The buckled beam consists of a hanging, elastic steel stripe, which can come to rest over either of two magnets that are fixed below it (see e.g. [86, 87]). The whole configuration (hanging stripe and magnets on a bottom plate) is driven periodically by a shaker.

On the other hand, Pohl's wheel consists of a wheel that is fixed via a spiral spring to its axis. A weight attached to the border of the wheel comes to rest on one side or the other. Damping is controlled by an electromagnetic brake. Like the buckled beam, this device is driven periodically by a shaker (search for "Pohl's wheel" on the Internet).

Both devices described above can be approximately described by Duffing's equation

$$m\frac{d^2x}{dt^2} = A\cos(\omega t) - c\frac{dx}{dt} - \alpha x^3 + \beta x\ , \tag{8.50}$$

where m is the mass and c is the damping constant. A and ω are the driving parameters. The force F corresponds to a two-well potential V, i.e. $F = -\frac{dV}{dx}$, where $V = \alpha\frac{x^4}{4} - \beta\frac{x^2}{2}$. The two minima of V correspond to the two points of rest in the devices described above. Chaos is obtained e.g. by setting $m = 1$, $A = 0.3$, $\omega = 1$, $c = 0.25$ and $\alpha = \beta = 1$.

Holmes' analysis [88] showed that a Poincaré section of solutions of Duffing's equation can be approximated by the cubic map

$$x_{n+1} = y_n\ , \tag{8.51}$$

$$y_{n+1} = -b\,x_n + r\,y_n - y^3\ . \tag{8.52}$$

Figures 8.26 and 8.27 show λ, as calculated with this map, for two different values of b and periodical modulations of r.

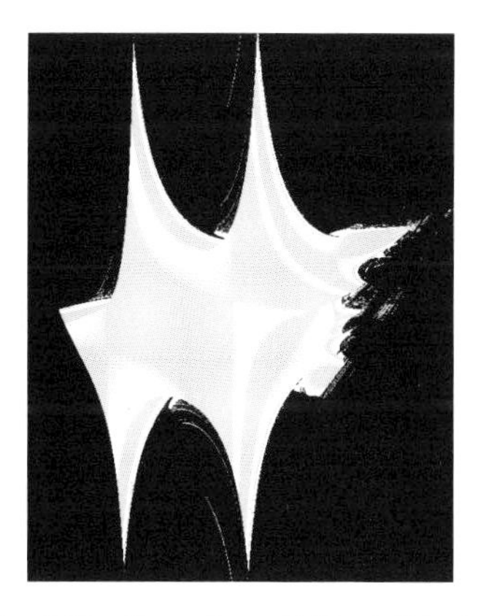

Fig. 8.26 Equations (8.51, 8.52): $b = 0.5$. B versus A. r:AAB AAB..., $n_{\text{prev}} = 500$, $n_{\max} = 2000$, $x_0 = y_0 = 0.4$. L-shading(min=−1), LL:(−1.6, −7), UL:(−1.6, 8), LR:(3.2, −7)

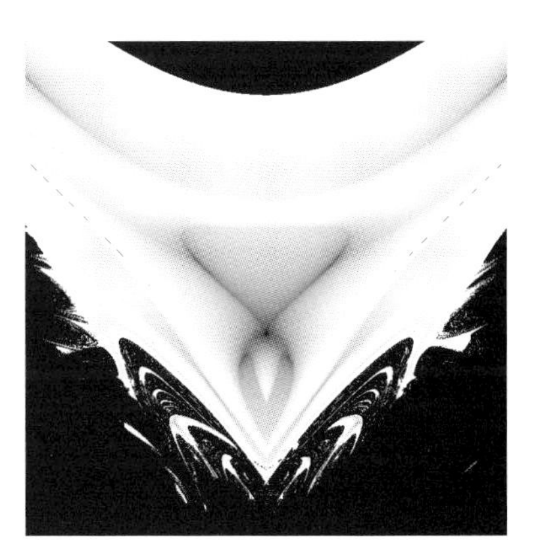

Fig. 8.27 Equations (8.51, 8.52): $b = 1.5$. B versus A. r:BA BA..., $n_{\text{prev}} = 1000$, $n_{\max} = 2000$, $x_0 = y_0 = 0.05$. L-shading. LL:(−1.92, 0.23), UL:(0.15, 2.3), LR:(0.23, −1.92)

Chapter 9

Maps of Generic Significance

9.1 The Gumowski-Mira attractors

In this section, graphical representations from the work of Igor Gumowski and Christian Mira from the University of Toulouse [89] will be presented. As an example, let us consider the map

$$x_{n+1} = y_n + b(1 - K\, y_n^2)y_n + F(x_n) \ , \tag{9.1}$$

$$y_{n+1} = -x_n + F(x_{n+1}) \ , \tag{9.2}$$

$$F(x) = rx + 2(1-r)\frac{x^2}{1+x^2} \ . \tag{9.3}$$

Figures 9.1 and 9.2 show, for two different values of K, λ on the plane defined by r (abscissa) and b (ordinate). Plate 4 ($K = 0.05$, $b = 0.005$ and $r = -0.495$) and Plate 5 ($K = 0.05$, $b = 0.006$ and $r = -0.899$) show colour representations of chaotic attractors obtained with this map. The colouring in these Plates was done "historically", meaning that the colour changes as the iterations proceed. Figure 9.3 shows λ for constant values of K and b, while r alternates as ABAB.... Plate 6 shows transients, as a chaotic attractor is approached. Here, $K = 0.05$, $b = 0.001$ and $r = 0.1$. After determining polar coordinates (r, ϕ) of each iterated point (setting the centre of the picture as the origin of the coordinates), r was plotted as ordinate and ϕ as abscissa on the plate. The calculations were done by starting at 1600 initial conditions, all corresponding to points lying equidistantly within an equilateral triangle in the centre of the plate. The iterates of these points were subsequently located within hill-shaped figures (distorted triangles) that became smaller and smaller and increasingly distorted, until they were "trapped" in the chaotic attractor that appears as a "sea" at the bottom of the plate.

Fig. 9.1 Equations (9.1, 9.2): $K = 0.05$. b versus r. $n_{\text{prev}} = 50$, $n_{\text{max}} = 300$, $x_0 = y_0 = -0.5$. L-shading. LL:$(-3.15, -1.6)$, UL:$(-3.15, 0.5)$, LR:$(2.3, -1.6)$

Fig. 9.2 Equations (9.1, 9.2): As Fig. 9.1, but $K = 10$. $n_{\text{prev}} = 100$, $n_{\text{max}} = 200$, $x_0 = y_0 = 0.5$. L-shading. LL:$(-0.5, -0.1)$, UL:$(-0.5, 1.05)$, LR:$(1.16, -0.1)$

A slow transient as the system approaches a single point is shown in Plate 7; as the iteration process goes on, the colour changes from red to yellow. Plate 8 shows a slow transient process as the system approaches an attractor consisting of four alternating points ($K = 0.05$, $b = 0.005$ and $r = 0.1$). Iterations were started from 800×600 equidistant points on the plane. As the iteration process goes on, the colour of the points changes from red to yellow, while they are represented (three-dimensionally) as increasing their height.

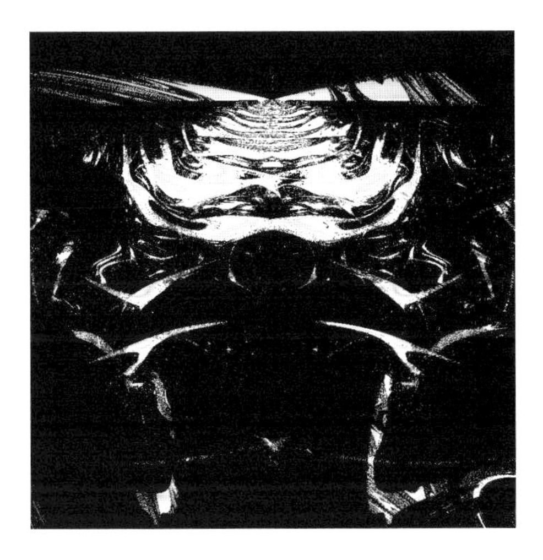

Fig. 9.3 Equations (9.1, 9.2): $K = 10$, $b = 0.6$. B versus A r:AB AB..., $n_{\text{prev}} = 100$, $n_{\max} = 200$, $x_0 = y_0 = 0.5$. L-shading. LL:(−1.5, 0.5), UL:(0.24, 2.24), LR:(0.5, −1.5)

9.2 The logistic equation

The equation

$$x_{n+1} = rx_n(1 - x_n) \tag{9.4}$$

may well have been treated in the preceding chapter on maps with applications, since it can be used to describe population dynamics. In that context, the factor r is the reproductive rate and the factor $(1 - x_n)$ describes the limitations on resources in a growing population. Besides its applications, however, the generic significance of this equation has been found to be so overwhelming that I chose to include it in this chapter.

Plates 9 through 19 show λ for the logistic equation using different A-B-sequences for r. The well-known period-doubling cascade at constant r (periods 2, 4, 8,...) leading to period ∞, i.e. to chaos, occurs for r =A=B, i.e. on the diagonal from the lower left to the upper right corner of Plate 9. Such a period doubling is one of the routes to chaos; moreover, it has universal numerical features found in all realms of science (see Cvitanović's book *Universality in Chaos* [90]). The reason for this universal behaviour is that many properties of the logistic equation are valid for any map $g : [a, b] \to [a, b]$ with $g(a) = g(b) = a$ and with one parabolic maximum in $[a, b]$. In fact, any such map is topologically conjugated to the logistic map, and thus exhibits analogous dynamical properties. A discussion of properties of the logistic equation in the light of diagrams, such as those shown in the colour plates here, was worked out mainly by Jaime Rössler (University of Chile) and myself [91–93]. As a particular feature, the period-3 window within the chaotic domain (tiny yellow spot at the upper right of Plate 9 for $3.284... < r < 3.8415...$) is shown enlarged on Plate 10, illustrating the occurrence of self-similarity, which will be discussed in Chapter 10.

Fig. 9.4 Equation (9.5): r versus α. $n_{\text{prev}} = 100$, $n_{\max} = 300$, $x_0 = 0.5$. D-shading. LL:(0, 0.65), UL:(0, 2.95), LR:(1, 0.65)

9.3 The discontinuous logistic equation

A dynamic behaviour that is dramatically different from that of the logistic equation (and of any map topologically conjugated with the logistic equation; see preceding section) occurs by perturbing the equation with a discontinuity at its maximum. One example of a map with such a discontinuity is

$$x_{n+1} = \begin{cases} 1 - r\,x_n^2 & \text{if} \quad x_n > 0 \\ \alpha - r\,x_n^2 & \text{else} \end{cases}, \tag{9.5}$$

which was studied in Refs. [94, 95]. This type of map can be obtained from appropriate Poincaré sections of the Lorenz model, which is well known to reflect dominant convective properties of fluids [96]. Figure 9.4 shows λ in the plane defined by α (abscissa) and r (ordinate). Alternating r as ABAB... yields Figs. 9.5 and 9.6 for two different values of α.

Fig. 9.5 Equation (9.5): $\alpha = 0.25$. B versus A r:AB AB..., $n_{\text{prev}} = 200$, $n_{\max} = 1000$, $x_0 = 0.5$. D-shading. LL:(0.912, 0), UL:(0.912, 1.1735), LR:(2.496, 0)

Introducing a transformation of variables in these equations, such that the interval [0, 1] is mapped into itself [8], leads to

$$x_{n+1} = \begin{cases} r\,x_n(1-x_n) & \text{if} \quad x_n > 0.5 \\ r\,x_n(1-x_n) + \frac{1}{4}(\alpha-1)(r-2) & \text{else} \end{cases} . \qquad (9.6)$$

Plates 20 through 24 show λ for these equations with different values of α and different A-B-sequences for r.

Fig. 9.6 As Fig. 9.5, but $\alpha = 0.4$, $n_{\text{prev}} = 100$, $n_{\text{max}} = 300$, $x_0 = 0.6$. D-shading. LL:(0.6, 1.375), UL:(1.375, 2.15), LR:(1.375, 0.6)

9.4 Discontinuity of the slope

In a celebrated paper in the journal *Nature*, which has been cited over 1800 times, Robert May published a variety of extremely simple mathematical models with very complicated dynamics [97]. One of these models is the following map, which is continuous, but has a discontinuous derivative

$$x_{n+1} = \begin{cases} b\,x_n & \text{for} \quad x_n \leq 1 \\ b\,x_n^{1-r} & \text{else} \end{cases} . \tag{9.7}$$

Alternating r as ABAB... and holding b constant yields Fig. 9.7.

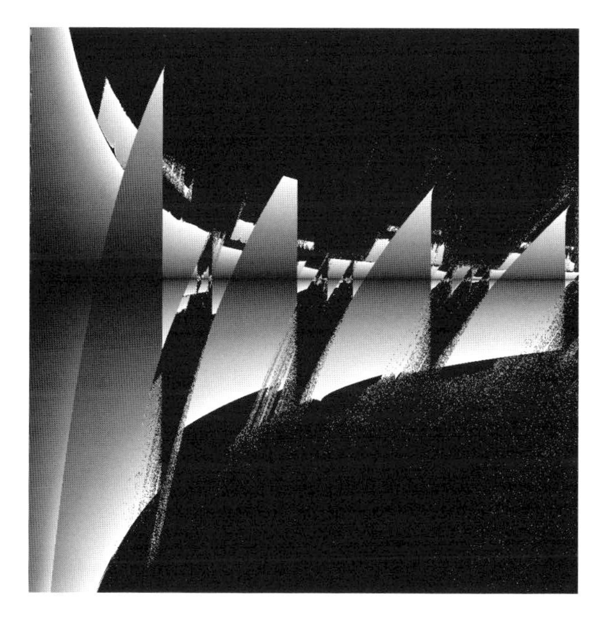

Fig. 9.7 Equation (9.7): $b = 50$. B versus A . r=AB AB..., $n_{\text{prev}} = 100$, $n_{\max} = 200$, $x_0 = 0.4$. D-shading. LL:(2, 0.5), UL:(2, 1.4), LR:(10.2, 0.5)

9.5 The KST-map

The following, one-humped, asymmetric, continuous map was proposed by Z. Kaufman, P. Szépfalusy and T. Tél (unpublished) and presented in Ref. [98]:

$$x_{n+1} = 1 + r\,|x_n|^b \text{sgn}(x_n) - a\,|x_n|^z . \tag{9.8}$$

Its name KST is formed from the authors' initials. Setting constant values for z and a, one obtains Fig. 9.8 on the plane defined by r (abscissa) and b (ordinate).

Fig. 9.8 Equation (9.8): $z = 2$, $a = 3$. b versus r. $n_{\text{prev}} = 100$, $n_{\text{max}} = 500$, $x_0 = 0.5$. D-shading. LL:(1.36, 1.23), UL:(1.36, 2), LR:(2.6, 1.23)

9.6 The periodically driven Poincaré oscillator

Probably the simplest possible continuous model of a limit cycle oscillator is the Poincaré oscillator

$$dr/dt = k\,r(1 - r)\ , \tag{9.9}$$

$$d\theta/dt = 2\pi\ , \tag{9.10}$$

where r and θ are polar coordinates (see [99]). We consider now periodic perturbations consisting of infinitely fast horizontal translations (on the phase plane) by an amount b. The effect of such a periodic stimulation in the limit $k \to \infty$ (very fast relaxation in the radial direction) is given by

$$\phi_{i+1} = g(\phi_i, b) + \tau , \tag{9.11}$$

where ϕ_i is the phase of the oscillator just before the ith stimulus, τ is the time between the stimuli relative to the period of the unperturbed oscillator (see [99]), and $g(\phi, b)$ is given by

$$\cos(2\pi g(\phi, b)) = (b + 2\pi\,\phi)/(1 + 2b\cos(2\pi\,\phi) + b^2)^{\frac{1}{2}} . \tag{9.12}$$

Figure 9.9 shows λ in the plane defined by b (abscissa) and $2\pi\tau$ (ordinate).

Fig. 9.9 Equations (9.11, 9.12): $2\pi\tau$ versus b. $n_{\rm prev} = 1000$, $n_{\rm max} = 1000$, $x_0 = 0.5$. D-shading. LL:(0.54, 1.16), UL:(1.02, 1.64), LR:(1.16, 0.54)

The stimulated Poincaré oscillator has been associated with some applications, such as periodically stimulated biological oscillators [100–104]. However, owing to its general features, it was included here in this chapter on generic maps. In one of the applications, namely in cardiology, every time ϕ passes zero, an action potential at the cell membrane is initiated; a horizontal translation to the right (left) on the phase plane corresponds roughly to a depolarizing (hyperpolarizing) stimulus of magnitude b.

9.7 The Hénon map

In Section 7.4, we derived the Hénon map [55], from a special case of the kicked rotator. Because of its generic properties, this map has become the "fruit fly" in investigations of two-dimensional maps; it is therefore treated in this chapter. By introducing simple transformations of parameters and variables in the equations derived in Section 7.4 one obtains the version

$$x_{n+1} = r - x_n^2 + b\, y_n \,, \tag{9.13}$$

$$y_{n+1} = x_n \,. \tag{9.14}$$

Figure 9.10 shows λ on the plane defined by r (abscissa) and b (ordinate).

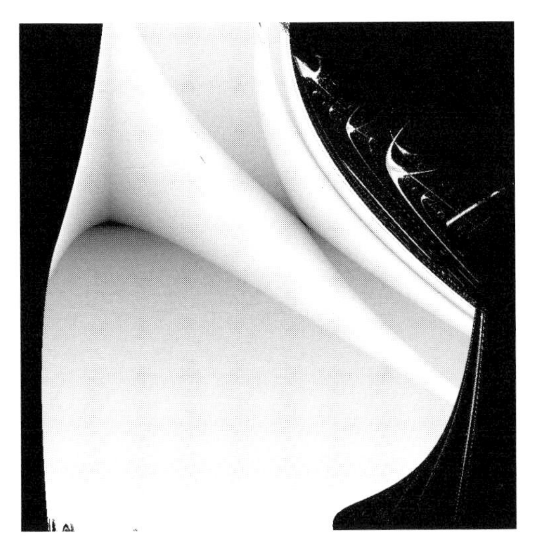

Fig. 9.10 Equations (9.13, 9.14): b versus r. $n_{\text{prev}} = 100$, $n_{\text{max}} = 2000$, $x_0 = y_0 = 0.4$. L-shading. LL:(−0.47, −0.97), UL:(−0.47, 0.63), LR:(2.1, −0.97)

9.8 The Hénon-Lozi map

In spite of the formal simplicity of the Hénon map (previous Section and Section 7.4), mathematical analyses have often proven difficult due to the quadratic term (see [105]). Due to Lozi [106] it is preferable to analyse a piecewise linear version of this map, which is called the Hénon-Lozi map:

$$x_{n+1} = 1 - r\,|x_n| + y_n \,, \tag{9.15}$$

$$y_{n+1} = b\, x_n \,. \tag{9.16}$$

Details can be found in Ref. [105]. Here, a λ diagram for b constant and r modulated as BABA... is presented (Fig. 9.11).

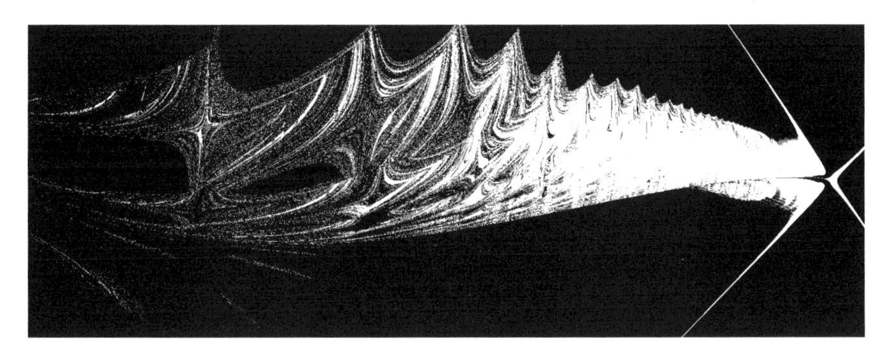

Fig. 9.11 Equations (9.15, 9.16): $b = 0.994$. B versus A . R:BA BA..., $n_{\text{prev}} = 100$, $n_{\text{max}} = 200$, $x_0 = y_0 = 0.4$. L-shading. LL:(1.483, 2.35), UL:(2.15, 1.794), LR:(−0.35, 0.15)

9.9 The Adams-Bashforth integration procedure

The simplest method for solving ordinary differential equations of the first order $\frac{dx}{dt} = g(x)$ is Euler's method, i.e. using the map $x_{n+1} = x_n + hg(x_n)$, where h is the time step. When using this method, only the previous value x_n is used to determine x_{n+1}, which is why it is called a "one-step" method. More accurate results are obtained by considering values x_{n-1}, x_{n-2}, etc. ("multistep" methods). A well-known two-step procedure is the Adams-Bashforth method, which reads:

$$x_{n+1} = x_n + \frac{h}{2}(3g(x_n) - g(x_{n-1})) \ . \tag{9.17}$$

Defining $x_{n-1} = y_n$, i.e. $x_n = y_{n+1}$, one obtains a two-dimensional discrete map:

$$x_{n+1} = x_n + \frac{h}{2}(3g(x_n) - g(y_n)) \ , \tag{9.18}$$

$$y_{n+1} = x_n \ . \tag{9.19}$$

As a differential equation, let us consider the continuous logistic equation $\frac{dx^*}{dt} = rx^*(1 - x^*/K)$, which describes a population of size x^*, r being the reproduction rate and K the carrying capacity of the medium. The normalization $x = x^*/K$ leads to $\frac{dx}{dt} = r\,x(1-x)$, which is set as $g(x)$ in Equations 9.18 and 9.19. Figure 9.12 shows λ for a constant value of h and modulating r as ABAB....

Fig. 9.12 Equations (9.18, 9.19): $h = 1$. B versus A r:AB AB..., $n_{\text{prev}} = 200$, $n_{\text{max}} = 200$, $x_0 = y_0 = 0.5$. L-shading. LL:(−4.7, 0), UL:(0, 4.7), LR:(0, −4.7)

9.10 Driven linear systems and the Degn-map

It is well known that exposing a non-linear system to periodical perturbations can cause chaotic oscillations. We have, in fact, encountered this phenomenon in many instances in this book. In contrast to the apparent requirement that the system must be non-linear, Degn [107, 108] found chaotic behaviour for a periodically perturbed linear system, namely

$$x_{n+1} = c\left(x_n - \frac{1}{2}\right) + \frac{1}{2} + R\sin(2\pi\, r\, y_n)\;, \tag{9.20}$$

$$y_{n+1} = (y_n + x_{n+1}) \bmod \frac{1}{b}\;. \tag{9.21}$$

In contrast to Degn's formulation, it was found convenient to introduce a modulo operation in Equation 9.21. Figure 9.13 shows the periodical and chaotic domains in the plane defined by r (abscissa) and b (ordinate) for $c = b$ and a constant R. Setting constant values for c, b and R while alternating r as BABA... yields Fig. 9.14.

Fig. 9.13 Equations (9.20, 9.21): $c = b$. $R = 0.1$. b versus r. $n_{\text{prev}} = 100$, $n_{\max} = 300$, $x_0 = y_0 = 0.4$. D-shading. LL:(0.2, −1.15), UL:(0.2, 1.15), LR:(2.9, −1.15)

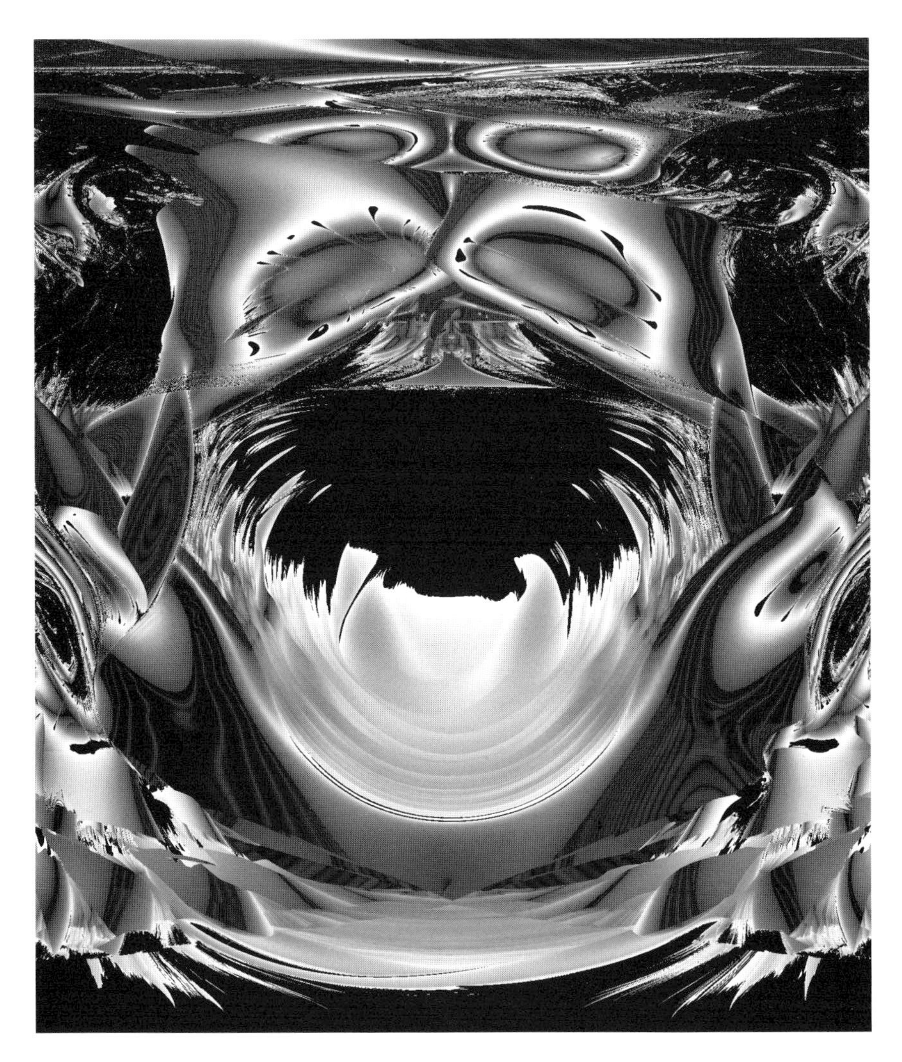

Fig. 9.14 Equations (9.20, 9.21): $b = 0.3$, $c = -0.8$, $R = 0.3$. B versus A. r:BA BA..., $n_{\text{prev}} = 100$, $n_{\text{max}} = 300$, $x_0 = y_0 = 0.4$. D-shading. LL:(0.38, 0.485), UL:(0.775, 0.88), LR:(0.485, 0.38)

9.11 Chaos via quasiperiodicity

The route to chaos via period-doubling was mentioned in previous sections. Another route consists in the breakdown of a system that is describable by incommensurate frequencies, i.e. the breakdown of quasiperiodicity. In this scenario, a torus in phase space (resulting from the quasiperiodical dynamics) wrinkles at all length scales and goes over into a chaotic attractor. Quasiperiodicity can be described by the circle map

$$x_{n+1} = x_n + \Omega - \left(\frac{r}{2\pi}\right) \sin(2\pi\, x_n) \ , \tag{9.22}$$

where x_{n+1} is the angle defined by the orbit on a Poincaré cut of the torus. If one intends to examine the breakdown of such a torus, then the equation above can be extended to include dynamics in the radial direction by introducing a variable y_n as follows:

$$x_{n+1} = x_n + \Omega - \left(\frac{r}{2\pi}\right) \sin(2\pi x_n) + b\, g(y_n) \ , \tag{9.23}$$

$$y_{n+1} = b\, g(y_n) - \left(\frac{r}{2\pi}\right) \sin(\pi\, x_n) \ . \tag{9.24}$$

Following Ref. [109], g is chosen here as the linear function $g(y_n) = y_n$.

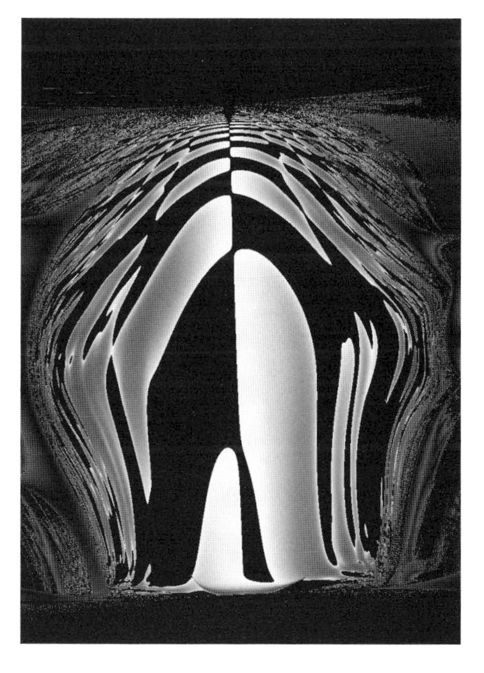

Fig. 9.15 Equations (9.23, 9.24): $\Omega = 0.58681$. b versus r. $n_{\text{prev}} = 100$, $n_{\max} = 300$, $x_0 = y_0 = 0.5$. D-shading. LL:(−1.1, −1.16), UL:(−1, 1.39), LR:(1.05, −1.24431)

Setting a constant value for Ω, one then obtains Fig. 9.15 for λ in the plane defined by r (abscissa) and b (ordinate). A small section of this diagram is shown in Fig. 9.16.

Owing to the fact that the trajectory remains very close to the torus,

Fig. 9.16 Equations (9.23, 9.24): Enlarged small section taken at the lower right of Fig. 9.15. $n_{\text{prev}} = 100$, $n_{\max} = 300$, $x_0 = y_0 = 0.5$. L-shading. LL:(0.59, −1.16), UL:(0.59, −0.34), LR:(0.79, −1.16)

λ is very close to zero. This entails very slow relaxations, i.e. very long transients. Figures 9.15 and 9.16 are therefore only valid for exactly 100 pre-iterations and 300 subsequent iterations. For any other number of iterations, details of these figures change, although they remain qualitatively similar. Convergence could not be achieved.

9.12 A map with an arbitrarily large number of coexisting attractors

Feudel *et al.* [110] reported on a map that yields an unusually large number of coexisting periodical attractors (100 or more). Their basins of attraction in phase space are strongly interwoven, so that the actual number of attractors that are determined depends on the density of the considered grid of initial conditions. The map reads

$$x_{n+1} = [x_n + y_n] \bmod (2\pi) \ , \tag{9.25}$$

$$y_{n+1} = (1-b)y_n + r \sin(x_n + y_n) \ . \tag{9.26}$$

For the particular initial conditions $x_0 = y_0 = 0.4$ one obtains λ in the parameter plane (abscissa r; ordinate b), as given by Fig. 9.17. Changing the initial conditions affects the lower part of the figure, especially at its right. It is, in fact, in this region where the many coexisting attractors occur. The rest of Fig. 9.17 is unaffected by the choice of initial conditions.

Fig. 9.17 Equations (9.25, 9.26): b versus r. $n_{\text{prev}} = 100$, $n_{\max} = 100$, $x_0 = y_0 = 0.4$. D-shading. LL:(0, 0), UL:(0, 2), LR:(5.1, 0)

9.13 The Maryland map

Grebogi *et al.* [111] reported the following map to demonstrate that the basins of attraction for different attractors can have fractal structures in phase space:

$$x_{n+1} = [x_n + b\,\sin(2x_n) - r\,\sin(4x_n) - y_n\,\sin(x_n)]\,\mathrm{mod}\,(2\pi)\ , \quad (9.27)$$

$$y_{n+1} = -J\,\cos(x_n)\ . \quad (9.28)$$

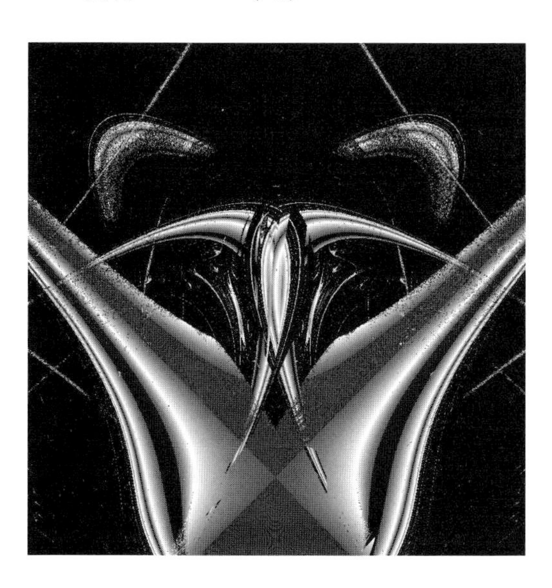

Fig. 9.18 Equations (9.27, 9.28): $J = 0.2$, $b = 1.99$. B versus A. r:BA BA..., $n_{\mathrm{prev}} = 200$, $n_{\mathrm{max}} = 200$, $x_0 = y_0 = 0.4$. D-shading. LL:(−1.55, −0.315), UL:(−0.315, 0.92), LR:(−0.315, −1.55)

Here we call this map the Maryland map, in consideration of the authors' affiliation to the University of Maryland. The map has two fixed points (x_n, y_n): $(0, -J)$ and (π, J). The initial conditions yielding a given fixed point, i.e. the basin of attraction of that fixed point, has a fractal structure, i.e. a highly interleaved structure that looks similar, no matter how small a subset one examines. In principle one may adapt the technique described in this book to the visualization of fractal basins. For this, the coordinates would have to be the initial conditions and grey shadings would have to be chosen, so as to distinguish the λ-values of the attractors attained for each pair of coordinates. This could be implemented in the future.

If r is modulated periodically as BABA..., the map given by Equations (9.27, 9.28) can become chaotic. Applying the technique described in this book, one then obtains the λ-diagram shown in Fig. 9.18. We will come back to these equations in Chapter 11.

9.14 Riddled and intermingled basins

Riddled or intermingled basins present more serious obstructions to predictability than the fractal basins mentioned in the previous section (see e.g. [112–116] and references within).

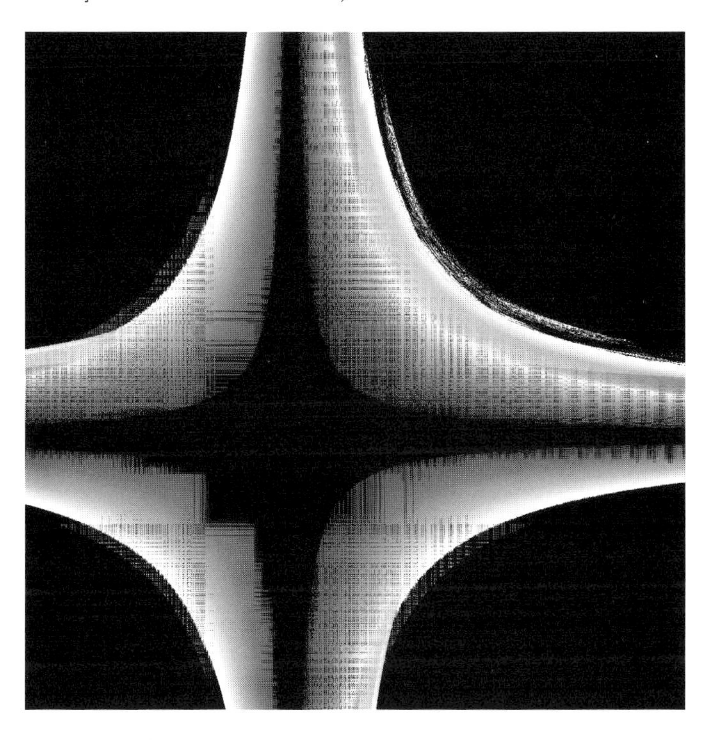

Fig. 9.19 Equations (9.29, 9.30): $c = 1.3$, $b = 0.42$. B versus A . r:AB AB..., $n_{\text{prev}} = 100$, $n_{\max} = 200$, $x_0 = y_0 = 0.1$. D-shading. LL:(−1, −1), UL:(−1, 2), LR: (2, −1)

Let us give a precise definition of the phenomenon: The basin β(A) of an attractor A is riddled to the basin β(B) of an attractor B if in any neighbourhood (in phase space) of a point in $\beta(A)$, i.e. an initial condition leading to A, there exists a point in β(B), i.e. of an initial condition leading to B. If, in addition, β(B) is riddled to β(A), then the basins are known as intermingled. Since the neighbourhood under consideration can be of any size, this means that improving the computational or measurement error is of no help for knowing if the system will end up in A or in B. In this section it will be demonstrated how this phenomenon can be visualized with λ-diagrams.

We consider the following piecewise linear map:

$$x_{n+1} = b\,x_n + \frac{c}{2}\left(1-\frac{b}{c}\right)\left[\left|x_n+\frac{1}{c}\right| - \left|x_n-\frac{1}{c}\right|\right] + r_1\,(y_n - x_n) \quad (9.29)$$

$$y_{n+1} = b\,y_n + \frac{c}{2}\left(1-\frac{b}{c}\right)\left[\left|y_n+\frac{1}{c}\right| - \left|y_n-\frac{1}{c}\right|\right] + r_2(y_n - x_n) \quad (9.30)$$

[117]. We set constant values for b and c. $r_1 = r_2 = r$ is modulated as ABAB....

Figure 9.19 shows λ in the set $[-1,2] \times [-1,2]$. Some regions in this figure give an indication of riddled basins, appearing here as a fabric-like structure. A much smaller square, namely $[0.7, 0.7000005] \times [0.7, 0.7000005]$, which is displayed in Fig. 9.20, is certainly of no help in deciding which initial conditions lead to chaos (black) and which to periodicity (white). In fact, the fabric-like texture remains unchanged for even smaller subsets. The basins are riddled to each other, i.e. intermingled.

Fig. 9.20 Equations (9.29, 9.30): Small enlarged section of Fig. 9.19. LL:(0.7, 0.7), UL:(0.7, 0.7000005), LR:(0.7000005, 0.7)

9.15 Transition between the tent map and the Bernoulli shift

The tent map and the Bernoulli shift are well known because, although they are very simple, they hold amazing properties. The tent map

$$x_{n+1} = \begin{cases} 2x_n & \text{for} \quad 0 < x_n \le \frac{1}{2} \\ 2 - 2x_n & \text{for} \quad \frac{1}{2} < x_n \le 1 \end{cases}, \quad (9.31)$$

yields a constant invariant measure, i.e. a constant probability distribution of x_n on the interval $[0, 1]$ for $n \to \infty$. In other words, it can be used as a random generator. (Note: in order to avoid numerical instabilities, one should use a parameter $2 - \epsilon$, ϵ being as small as possible, instead of the parameter 2 in Eq. 9.31.)

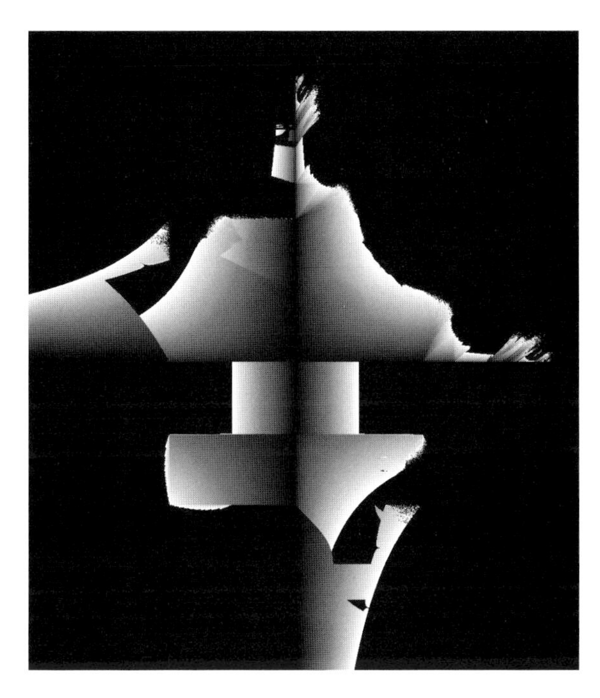

Fig. 9.21 Equation (9.33): B versus A. r:AB AB... $n_{\text{prev}} = 100$, $n_{\max} = 200$, $x_0 = 0.45$. D-shading. LL:(−0.03, −0.04), UL:(−0.03, 1.08), LR:(1.06, −0.04)

The other map, the Bernoulli shift, defined in the interval [0,1] as

$$x_{n+1} = [2x_n] \bmod 1 \ , \tag{9.32}$$

was given its name because it shifts binary digits to the left, while annihilating the first digit. For example, writing the numbers in the interval $[0, 1]$ in binary code, then $x_n = 0.1101110...$ implies $x_{n+1} = 0.101110....$ This means that rational initial conditions, e.g. 0.101010... or 0.1011000000..., lead to periodicity, while irrational initial conditions, i.e. aperiodic 0s and 1s, lead to chaos. In other words, the basins of attraction of periodicity and chaos are riddled to each other, i.e. intermingled (see previous section). This means that in any neighbourhood of an initial condition leading to chaos there exist initial conditions leading to periodicity, and *vice versa*.

Lima *et al.* [118] examined a map that displays a transition from the tent map to the Bernoulli shift:

$$x_{n+1} = \begin{cases} 2x_n & \text{for} \quad 0 \le x_n \le \frac{1}{2} \\ (4r-2)x_n + (2-3r) & \text{for} \quad \frac{1}{2} < x_n \le 1 \end{cases} . \tag{9.33}$$

The transition is obtained by changing the parameter r. The map is discontinuous at $x_n = \frac{1}{2}$, except for the case $r = 0$, which corresponds to the tent map. For $r = 1$ one gets the Bernoulli shift.

Figures 9.21 and 9.22 show λ for different periodical modulations of r.

Fig. 9.22 As Fig. 9.21, but r:A^{8}B^8 A^{8}B^8... LL:(−0.25, 0.51), UL:(0.45, 1.21), LR:(0.51, −0.25)

9.16 The Kaplan-Yorke map and fractal basin boundaries

The following, apparently simple but highly surprising map was studied by Kaplan and Yorke [119] and by McDonald *et al.* [120]:

$$x_{n+1} = [r\,x_n] \bmod 1 \,, \tag{9.34}$$

$$y_{n+1} = b\,y_n + \cos(2\pi x_n) \,. \tag{9.35}$$

Figure 9.23 shows λ for constant b and periodically modulated r.

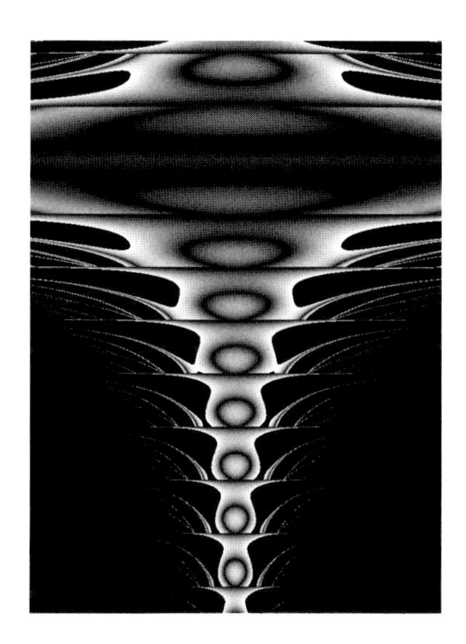

Fig. 9.23 Equations (9.34, 9.35): $b = 0.6$. r:BA BA..., $n_{\text{prev}} = 1$, $n_{\max} = 5$, $x_0 = 0.7$, $y_0 = 0.5$. D-shading. LL:(−0.5, −12.15), UL:(−0.5, 3.2), LR:(0.5, −12.15)

One obtains periodicity and chaos both for the low n considered in this figure, and for $n \to \infty$. Under different conditions, namely constant $r > b > 1$ it was shown [120] that periodicity and chaos disappear, while only two divergent attractors ($y_n \to -\infty$ and $y_n \to \infty$) are present. Although our graphical method is (in its present form) not adequate to describe this situation, the reader might be interested to know that the boundary between the basin for $+\infty$ and that for $-\infty$ is a fractal. Another unusual fact is that this fractal can be expressed analytically:

$$y_0 = -\sum_{j=1}^{\infty} b^{-j} \cos(2\pi\, r^{j-1} x_0) \,, \tag{9.36}$$

which is a continuous, but nowhere differentiable curve. The fractal dimension of this curve is given by $d = 2 - \frac{\ln b}{\ln r}$.

9.17 Chaos control via phase space compression

A simple method to control chaos is by defining subsets in phase space within which the system is forced to assume a given, constant value. Based on the proposal of Zhang and Shen [121], we use the following procedure for controlling chaos described by 1D maps $x_{n+1} = f(x_n)$:

$$x_{n+1} = \begin{cases} f(x_n) & \text{for} \quad x_{\min} < x_n < x_{\max} \\ x_{\max} & \text{for} \quad x_n \geq x_{\max} \\ x_{\min} & \text{for} \quad x_n \leq x_{\min} \end{cases} . \tag{9.37}$$

Here we choose the logistic map $f(x) = rx(1-x)$, setting $r = 4$, for which this map is chaotic. Figure 9.24, shows λ for $x_{\min} = 0$, $x_{\max} = r$, where r is modulated as ABAB.... The diagonal defined by r=A=B appears black because of superstability ($\lambda \to -\infty$); this is because the system gets stuck on the constant $x_{\min}$. The other black subset, namely that on the heart-shaped region, also indicates high stability. In general, our graphical analysis demonstrates success of chaos control.

Fig. 9.24 Equation (9.37): B versus A. r:AB AB..., $n_{\text{prev}} = 200$, $n_{\max} = 200$, $x_0 = 0.5$. D-shading. LL:(0.75, 0.75), UL:(0.75, 1), LR:(1, 0.75)

9.18 The Mandelbrot set

The reader may be surprised to find the well-known Mandelbrot set M, which has become an icon in non-linear science and is seemingly unrelated to the subject of this book, as a λ-diagram in Fig. 9.25.

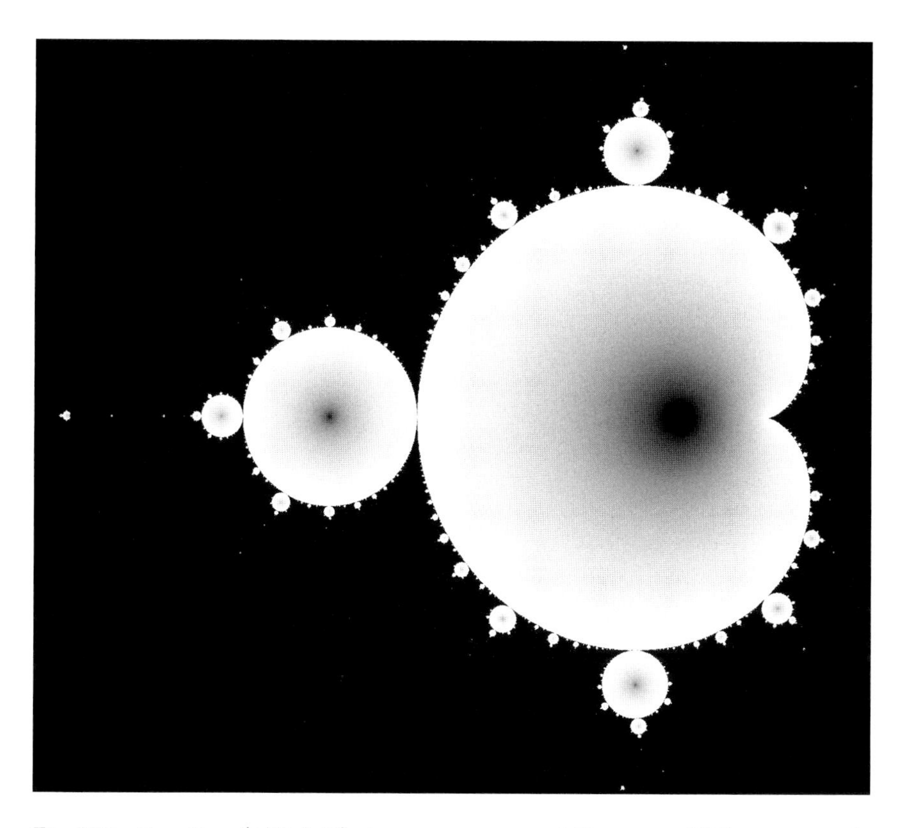

Fig. 9.25 Equations (9.39, 9.40): b versus r. $n_{\rm prev} = 100$, $n_{\rm max} = 1000$, $x_0 = y_0 = 0$. L-shading. LL:(−1.85, −1.05), UL:(−1.85, 1.05), LR:(0.55, −1.05)

For the construction of M, one considers the complex logistic equation

$$z_{n+1} = z_n^2 + c \ , \tag{9.38}$$

where c and z_n are complex numbers. Separating the real and the imaginary parts, this equation can be written as the two-dimensional map given by

$$x_{n+1} = x_n^2 - y_n^2 + r \ , \tag{9.39}$$

$$y_{n+1} = 2x_n \, y_n + b \ , \tag{9.40}$$

where $z_n = x_n + iy_n$ and $c = r + ib$. Now one should recall that M is defined on the complex parameter plane by the set of all c, such that the iteration starting at z_0 does not diverge, but leads to a periodic cycle [122]. Divergence (not chaos in the present case) leads to $\lambda > 0$, while periodical orbits lead to $\lambda < 0$, so that M is obtained.

A word of caution must be written here. The most exciting features of M occur at its border, especially in the regions where the subsets of M are tangential to each other: "seahorses", spiral structures and smaller sets similar to M appear along with other aesthetically attractive patterns [4]. The problem in reproducing these patterns with a λ-diagram is that in these regions (transitions between periodicity and divergence), $\lambda \approx 0$. This means that the so-called "critical slowing down" occurs, i.e. perturbations relax very slowly. In fact, they relax proportionally to $e^{\lambda n}$, so that as λ approaches zero, the n needed for relaxation approaches ∞. Therefore, the difference between the initial conditions and the attained attractors (for $\lambda \approx 0$) diminishes so slowly that it may take prohibitively long computing times to attain convergence. At reasonable computing times, the patterns at these borders may thus appear blurred.

9.19 Maps related to strange nonchaotic attractors

Many people who are familiar with, but not experts on chaos, often speak of chaos and fractals in the same breath. They are right, in part, as many chaotic phenomena are related to fractals; we have seen this e.g. in Section 7.4. However, the two concepts do not always overlap. For a better understanding, let us remember the significance of the two concepts. A fractal is a geometrical object that looks similar at all length scales and that is describable by a non-integer dimension. On the other hand, chaos is a temporal behaviour in which neighbouring initial conditions diverge in the average ("butterfly effect", $\lambda_{\max} > 0$). If a system does not dissipate energy, as is (mostly) the case for astronomical objects and (mostly) also in particle accelerators, the points in the chaotic regions of phase space are homogeneously distributed, i.e. far from being fractals. On the other hand, chaotic trajectories in dissipative systems dwell (for $t \to \infty$ or $n \to \infty$) in an attractor, i.e. in a set that attracts nearby trajectories; these attractors are fractals and are therefore called "strange" (see Section 7.4).

The surprise presented in this section is the existence of attractors that are strange but are not chaotic. Strange nonchaotic attractors can be obtained by quasiperiodically forcing a 2D map, such as the circle map (see [123]):

$$x_{n+1} = [x_n + 2\pi K + b\,\sin(x_n) + r\,\cos(y_n)] \bmod (2\pi)\ , \tag{9.41}$$

$$y_{n+1} = [y_n + 2\pi\,\omega] \bmod (2\pi)\ , \tag{9.42}$$

Fig. 9.26 Equations (9.41, 9.42): $K = 0.28$, $\omega = \frac{\sqrt{5}-1}{2}$. b versus r. $n_{\text{prev}} = 100$, $n_{\max} = 300$, $x_0 = 0.1$, $y_0 = 0.5$. D-shading. LL:(−10, −2.9), UL:(−10, 2.9), LR:(10, −2.9)

where ω is irrational. Setting constant values for K and $\omega = \omega_g = \frac{\sqrt{5}-1}{2}$, one obtains Fig. 9.26, which shows λ in the plane defined by r (abscissa) and b (ordinate). λ in a different parameter plane, namely that defined by K (abscissa) and r (ordinate), setting $b = 1.075$ and $\omega = \omega_g$, is shown in Fig. 9.27. The effect of modulating r as BABA... at $b = 1.075$, $\omega = \omega_g$ and $K = 0.28$ is shown in Fig. 9.28. An alternative way to generate strange nonchaotic attractors is the quasiperiodically forcing of the logistic map:

$$x_{n+1} = r[1 + b\cos(2\pi y_n)]x_n(1 - x_n) , \tag{9.43}$$

$$y_{n+1} = [y_n + \omega] \bmod 1 . \tag{9.44}$$

(See [124].) We set again $\omega = \omega_g$, hold $b = 0.2$ and alternate r as BABA..., obtaining Fig. 9.29.

Fig. 9.27 Equations (9.41, 9.42): $\omega = \frac{\sqrt{5}-1}{2}$, $b = 1.075$. r versus K. $n_{\text{prev}} = 100$, $n_{\max} = 300$, $x_0 = 0.1$, $y_0 = 0.5$. D-shading. LL:(−0.7, −3.9), UL:(−0.7, 3.9), LR:(0.7, −3.9)

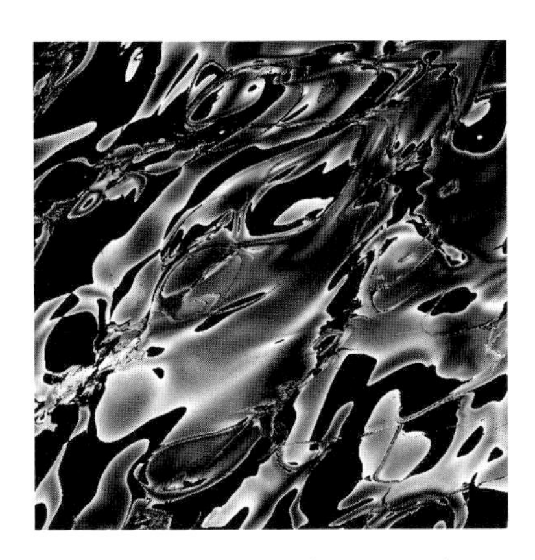

Fig. 9.28 Equations (9.41, 9.42): As Fig. 9.27, but $K = 0.28$. B versus A. r:BA BA..., $n_{\text{prev}} = 100$, $n_{\max} = 200$, $x_0 = y_0 = 0.4$. D-shading. LL:(−2, −6), UL:(−2, 12), LR:(13, −6)

Fig. 9.29 Equations (9.43, 9.44): $\omega = \frac{(\sqrt{5}-1)}{2}$, $b = 0.2$. B versus A. r:BA BA..., $n_{\text{prev}} = 100$, $n_{\max} = 200$, $x_0 = y_0 = 0.4$. D-shading. LL:(2.87, 2.87), UL:(2.87, 3.2), LR:(3.28, 2.87)

Yet another generator of strange nonchaotic attractors is

$$x_{n+1} = [x_n + 2\pi\,\omega] \bmod (2\pi) \quad , \tag{9.45}$$

$$y_{n+1} = \left(\frac{1}{2\pi}\right)(r\,\cos x_n + b)\sin(2\pi y_n)\ . \tag{9.46}$$

(See [125].) Now we write $\omega = \sqrt{2} - 1$; λ in the b-r-plane is shown in Fig. 9.30.

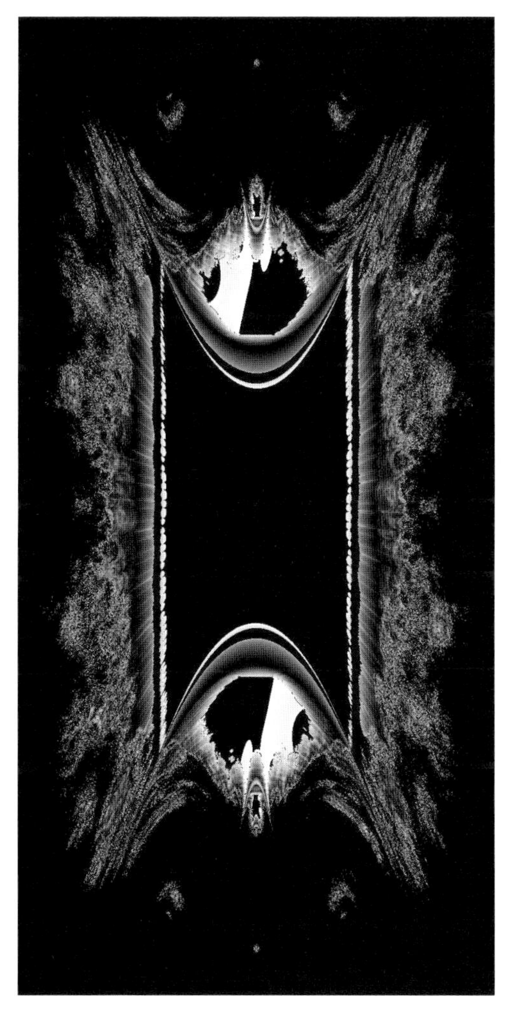

Fig. 9.30 Equations (9.45, 9.46): $\omega = \sqrt{2} - 1$. b versus r. $n_{\rm prev} = 100$, $n_{\rm max} = 300$, $x_0 = y_0 = 0.5$. D-shading. LL:(−5, −4), UL:(−5, 4), LR:(5, −4)

It is important to stress that for Figs. 9.26 through 9.30 a similar situation to that discussed in Section 9.11 arises again: these figures are transitory. This is because the attractors are nonchaotic; thus $\lambda = 0$ and the relaxation from the initial conditions to the attractor occurs at $n \to \infty$. As n increases, general features of these diagrams do not change substantially, but details do change. The figures show small fluctuations of λ around $\lambda = 0$. These fluctuations depend on the map and therefore the qualitative features of the figures shown here are not arbitrary.

9.20 The $\sin^2$-map

The map

$$x_{n+1} = b \sin^2(x_n + r) \ , \tag{9.47}$$

leading to Plate 25 through Plate 34, was derived in Section 8.14, i.e. in the chapter on maps with applications, as an approximation of the dynamics of an optical bistable liquid crystal [76–79]. Nevertheless, it is considered in the present chapter on generic systems because of two reasons: a) It has a large number of general properties; and b) the approximations to a physical system are too rough to stand up alone.

9.21 Discontinuous $\sin^2$-maps

We consider the map

$$x_{n+1} = \begin{cases} b \sin^2(x_n + r^m) + \alpha\, r^k & \text{if}\, [x_n + r^n] \bmod (\gamma\pi) \ < \ \mu\frac{\pi}{2} \\ b \sin^2(x_n + r^m) + \beta\, r^k & \text{else} \end{cases} . \tag{9.48}$$

Figure 9.31 shows λ for this map for $b = 2$, $k = 1$ and $\beta = 0$, the coordinates being r (abscissa) and α (ordinate). The map 9.48 is astonishingly fruitful in rendering different λ-diagrams upon alternation of r between two values A and B, as illustrated in Figs. 9.32 through 9.115, as well as in Plates 35–44.

Note that setting $m = n$ and $\gamma = \mu = 1$, the $\sin^2$-function is shifted vertically every half-period. For $m = n$ and $\gamma = \mu = 2$, the $\sin^2$-function is shifted vertically every period. For $m = n$, $\gamma = 2$ and $\mu = 1$ this shift occurs every $1\frac{1}{2}$ periods. In all cases, discontinuities arise at extrema, similarly to the simpler case of the discontinuous logistic equation (Section 9.3).

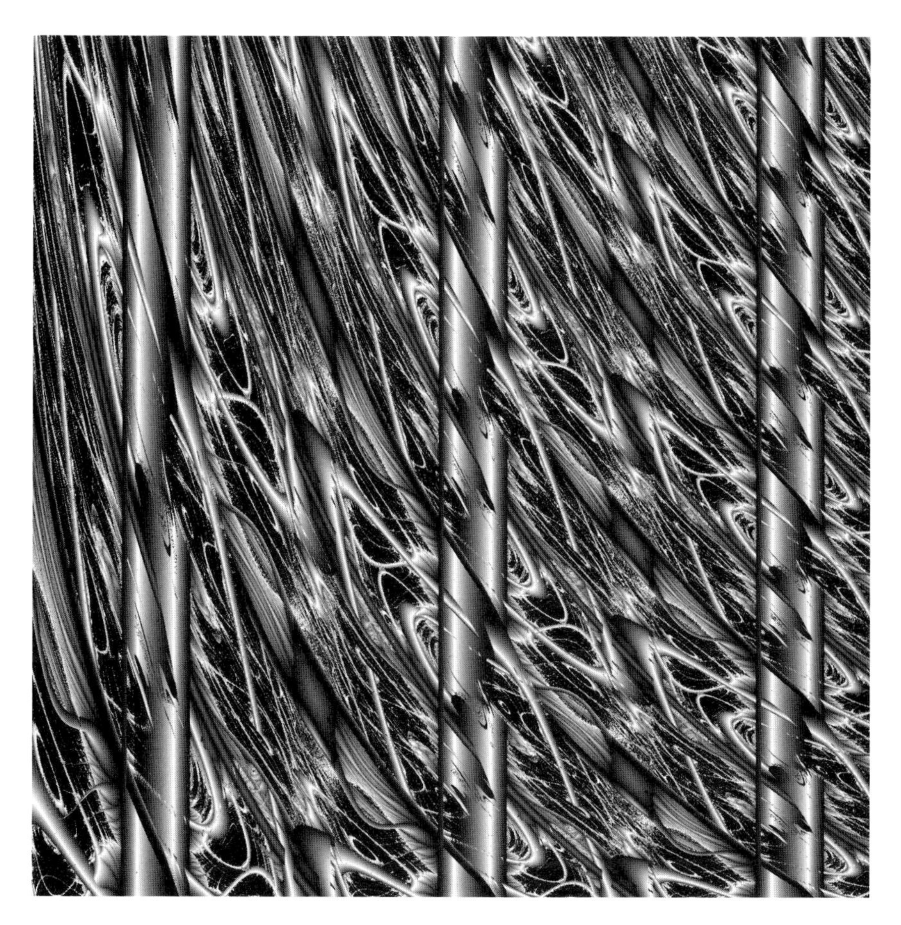

Fig. 9.31 Equation (9.48): $k = 1$, $b = 2$, $m = n = \mu = \gamma = 1$, $\beta = 0$. α versus r. $n_{\text{prev}} = 200$, n_{max}=500, $x_0 = 0.5$. D-shading. LL:(1.7, 0), UL:(1.7, 1.5), LR:(10, 0)

9.22 Other maps

The following maps demonstrate the diversity of λ-diagrams obtained by varying the mathematical formulations of the maps. Here, mainly exponential and trigonometric functions are included. Figures 9.116 through 9.159 show λ for these maps, both on r-b-planes and on A-B-planes.

Fig. 9.116:

$$x_{n+1} = [b\ \cosh(rx_n)]\mathrm{mod}(2b) \tag{9.49}$$

Fig. 9.117:

$$x_{n+1} = [\cosh(rx_n)]\mathrm{mod}\left(\frac{2}{b}\right) \tag{9.50}$$

Fig. 9.118:

$$x_{n+1} = b\,r\,e^{\sin(1-x_n)^3 \cos(-(x_n-r)^2)} - 1 \tag{9.51}$$

Fig. 9.119:

$$x_{n+1} = b\ \sin[(x_n - r)^3]\,e^{-(x_n-r)^2} \tag{9.52}$$

Fig. 9.120:

$$x_{n+1} = b\ \sin^4(x_n - r) \tag{9.53}$$

Fig. 9.121:

$$x_{n+1} = \cos(x_n + r)\cos(1 - x_n) \tag{9.54}$$

Fig. 9.122:

$$x_{n+1} = b\,(x_n - 1)^2 \sin^2(r - x_n) \tag{9.55}$$

Fig. 9.123:

$$x_{n+1} = b\,r\,e^{\cos^3(1-x_n)\sin^2(\pi-x_n)} \tag{9.56}$$

Fig. 9.124:

$$x_{n+1} = b\ \sin(x_n)\sin(r_n x_n) \tag{9.57}$$

Figs. 9.125 and 9.128:

$$x_{n+1} = (b + r)e^{\sin(1-x_n)^3 \cos(-(x_n-r)^2)} \tag{9.58}$$

Fig. 9.126, 9.152 and 9.158:

$$x_{n+1} = (b + r)e^{\sin(1-x_n)^3 \cos(-(x_n-r)^2)} - 1 \tag{9.59}$$

Fig. 9.127:

$$x_{n+1} = b\,e^{\sin((1-x_n)^3)}e^{\cos(-(x_n-r)^2)} \tag{9.60}$$

Fig. 9.129:

$$x_{n+1} = b\ \cos[e^{-(x_n-r)^2}] \tag{9.61}$$

Fig. 9.130:

$$x_{n+1} = b\,r^2 e^{\sin[(1-x_n)^3]} - 1 \tag{9.62}$$

Fig. 9.131:

$$x_{n+1} = b\,e^{\sin^3(1-x_n)} + r \tag{9.63}$$

Fig. 9.132:

$$x_{n+1} = r\,e^{-(x_n-b)^2} \tag{9.64}$$

Fig. 9.133:

$$x_{n+1} = b\,e^{\sin(r\,x_n)} \tag{9.65}$$

Fig. 9.134:

$$x_{n+1} = |b^2 - (x_n - r)^2|^{\frac{1}{2}} + 1 \tag{9.66}$$

Fig. 9.135:

$$x_{n+1} = [b + \sin(r\,x_n)^2]^{-1} \tag{9.67}$$

Fig. 9.136:

$$x_{n+1} = b\,e^{r(\sin(x_n)+\cos(x_n))^{-1}} \tag{9.68}$$

Fig. 9.137:

$$x_{n+1} = b\,(x_n - r)e^{-(x_n-r)^3} \tag{9.69}$$

Fig. 9.138:

$$x_{n+1} = b\,e^{\cos(1-x_n)\sin(\frac{\pi}{2})+\sin(r)} \tag{9.70}$$

Fig. 9.139:

$$x_{n+1} = b\,r\,e^{\sin(x_n-r)^4} \tag{9.71}$$

Fig. 9.140:

$$x_{n+1} = b\,r\,e^{\sin(1-x_n)^3} \tag{9.72}$$

Fig. 9.141:

$$x_{n+1} = b\,r\,\sin^2(b\,x_n + r^2)\cos^2(b\,x_n - r^2) \tag{9.73}$$

Fig. 9.142:

$$x_{n+1} = |r^2 - (x_n - b)^2|^{\frac{1}{2}} \tag{9.74}$$

Fig. 9.143:

$$x_{n+1} = b\,\cos(x_n - r)\sin(x_n + r) \tag{9.75}$$

Fig. 9.144:

$$x_{n+1} = (x_n - r)\sin[(x_n - b)^2] \tag{9.76}$$

Figs. 9.145 and 9.146:

$$x_{n+1} = \begin{cases} r\sin(\pi\,r)\sin(\pi\,x_n) & \text{if} \quad x_n > 0.5 \\ b\,r\sin(\pi r)\sin(\pi x_n) & \text{else} \end{cases} \tag{9.77}$$

Fig. 9.147:

$$x_{n+1} = r\sin(\pi\,r)\sin[\pi(x_n - b)] \tag{9.78}$$

Figs. 9.148 and 9.149:

$$x_{n+1} = b\,r\sin^2(b\,x_n + r^2)\cos^2((b\,x_n - r^2) - r \tag{9.79}$$

Figs. 9.150 and 9.151:

$$x_{n+1} = b\,r\sin^2(b\,x_n + r^2)\cos^2((b\,x_n - r^2) - 1 \tag{9.80}$$

Figs. 9.153 and 9.154:

$$x_{n+1} = b[2 + \sin((x_n \bmod 1) - r)]^{-1} \tag{9.81}$$

Figs. 9.155 and 9.156:

$$x_{n+1} = b\,r\exp\{\exp\{\exp\{x_n^3\}\}\} \tag{9.82}$$

Fig. 9.157:

$$x_{n+1} = b\,r e^{\sin^4(1-x_n^2)} \tag{9.83}$$

Fig. 9.159:

$$x_{n+1} = r(\sin(x_n) + b\sin(9x_n)) \tag{9.84}$$

Fig. 9.160:

$$x_{n+1} = b\,e^{\tan(r\,x_n) - x_n} \tag{9.85}$$

Fig. 9.161:

$$x_{n+1} = b\,e^{\cos(x_n^3\,r - b) - r} \tag{9.86}$$

Fig. 9.32 Equation (9.48): $b = 2.25$, $k = 2$, $m = n = \gamma = \mu = 1$, $\alpha = -1$, $\beta = 0$, r:AABB AABB..., $n_{\text{prev}} = 200$, $n_{\max} = 200$, $x_0 = 0.5$. D-shading. LL:(4.4348, 4.5432), UL:(4.5546, 4.663), LR:(4.5522, 4.4258)

Fig. 9.33 Equation (9.48): $b = 1.5$, $k = 2$, $m = n = \gamma = \mu = 1$, $\alpha = 0.6$, $\beta = 0$, B versus A, r:AABB AABB..., $n_{\text{prev}} = 100$, $n_{\max} = 200$, $x_0 = 0.5$. D-shading. LL:(5.13, 5.375), UL:(5.433, 5.678), LR:(5.375, 5.13)

Fig. 9.34 Equation (9.48): $b = 2$, $k = 2$, $m = n = \gamma = \mu = 1$, $\alpha = 2.4$, $\beta = 0$, B versus A, r:AABB AABB..., $n_{\text{prev}} = 200$, $n_{\max} = 200$, $x_0 = 0.5$. D-shading. LL:(0.796, 0.966), UL:(0.918, 1.088), LR:(0.966, 0.796)

Fig. 9.35 Equation (9.48): $b = 2$, $k = 2$, $m = n = \gamma = \mu = 1$, $\alpha = -0.4$, $\beta = 0$, B versus A, r:AABB AABB..., $n_{\text{prev}} = 200$, $n_{\max} = 200$, $x_0 = 0.3$. D-shading. LL:(5.5113, 5.5933), UL:(5.615, 5.697), LR:(5.5933, 5.5113)

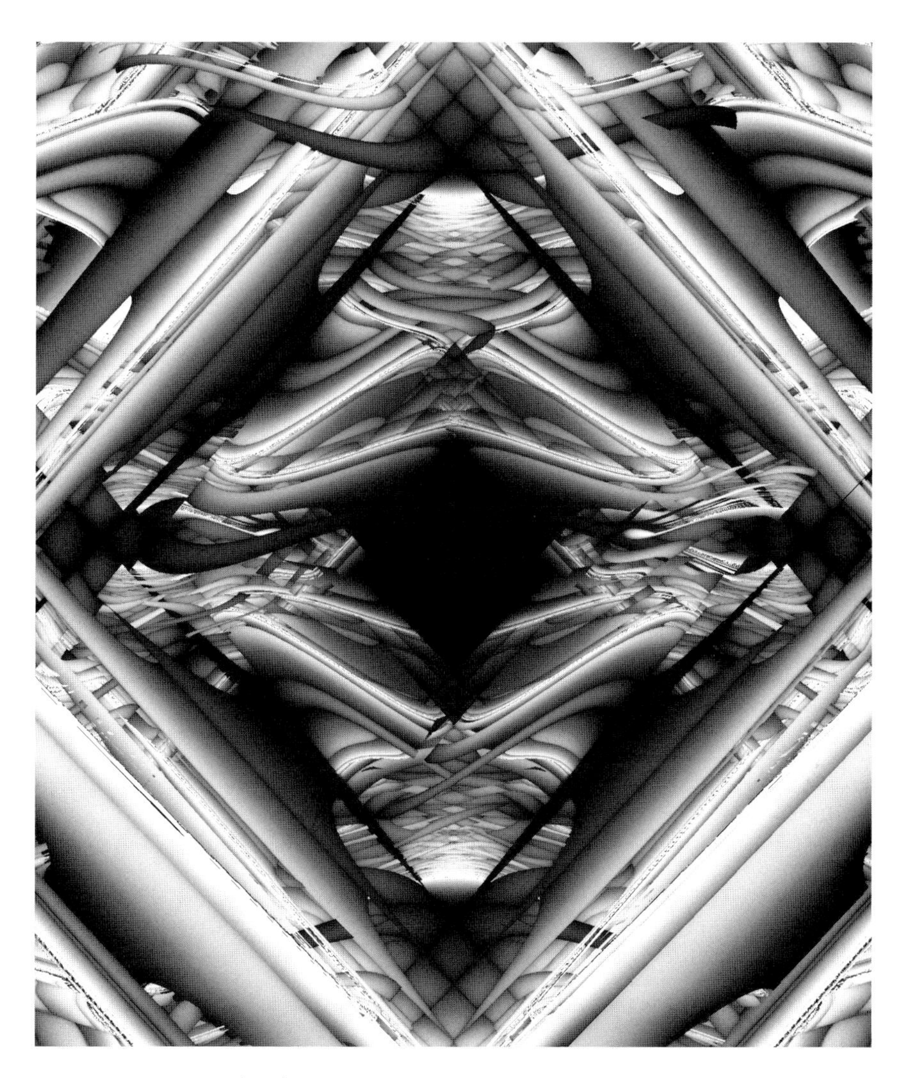

Fig. 9.36 Equation (9.48): $b = 1.5$, $k = 2$, $m = n = \gamma = \mu = 1$, $\alpha = -1.25$, $\beta = 0$, B versus A, r:AABB AABB..., $n_{\text{prev}} = 100$, $n_{\max} = 200$, $x_0 = 0.6$. D-shading. LL:(4.382, 4.733), UL:(4.771, 5.122), LR:(4.732, 4.383)

Fig. 9.37 Equation (9.48): $b = 2$, $k = 2$, $m = n = \gamma = \mu = 1$, $\alpha = -0.2$, $\beta = 0$, B versus A, r:AAABBB AAABBB..., $n_{\text{prev}} = 2000$, $n_{\text{max}} = 2000$, $x_0 = 0.5$. D-shading. LL:(4.569, 4.67), UL:(4.661, 4.762), LR:(4.67, 4.569)

Fig. 9.38 Equation (9.48): $b = 2$, $k = 2$, $m = n = \gamma = \mu = 1$, $\alpha = 0.7$, $\beta = 0$, B versus A, r:AABB AABB..., $n_{\text{prev}} = 200$, $n_{\text{max}} = 200$, $x_0 = 0.5$. D-shading. LL:(2.33, 1), UL:(2.33, 1.77), LR:(3.61, 1)

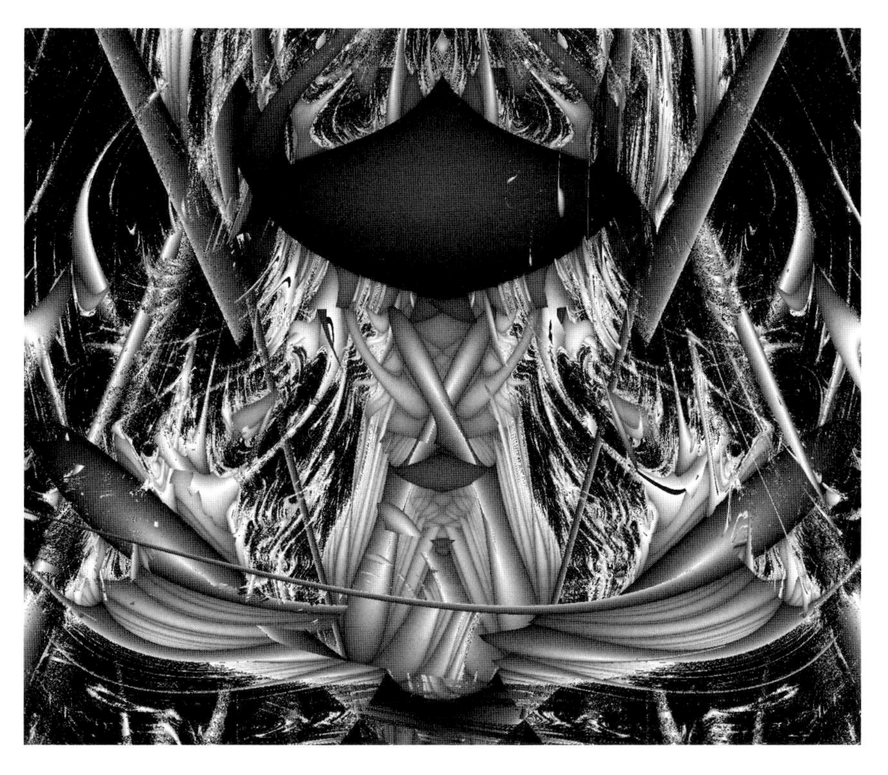

Fig. 9.39 Equation (9.48): $b = 2$, $k = 2$, $m = n = \gamma = \mu = 1$, $\alpha = 0.7$, $\beta = 0$, B versus A, r:AABB AABB..., $n_{\text{prev}} = 2000$, $n_{\text{max}} = 2000$, $x_0 = 0.3$. D-shading. LL:(5.009, 5.123), UL:(5.059, 5.173), LR:(5.123, 5.009)

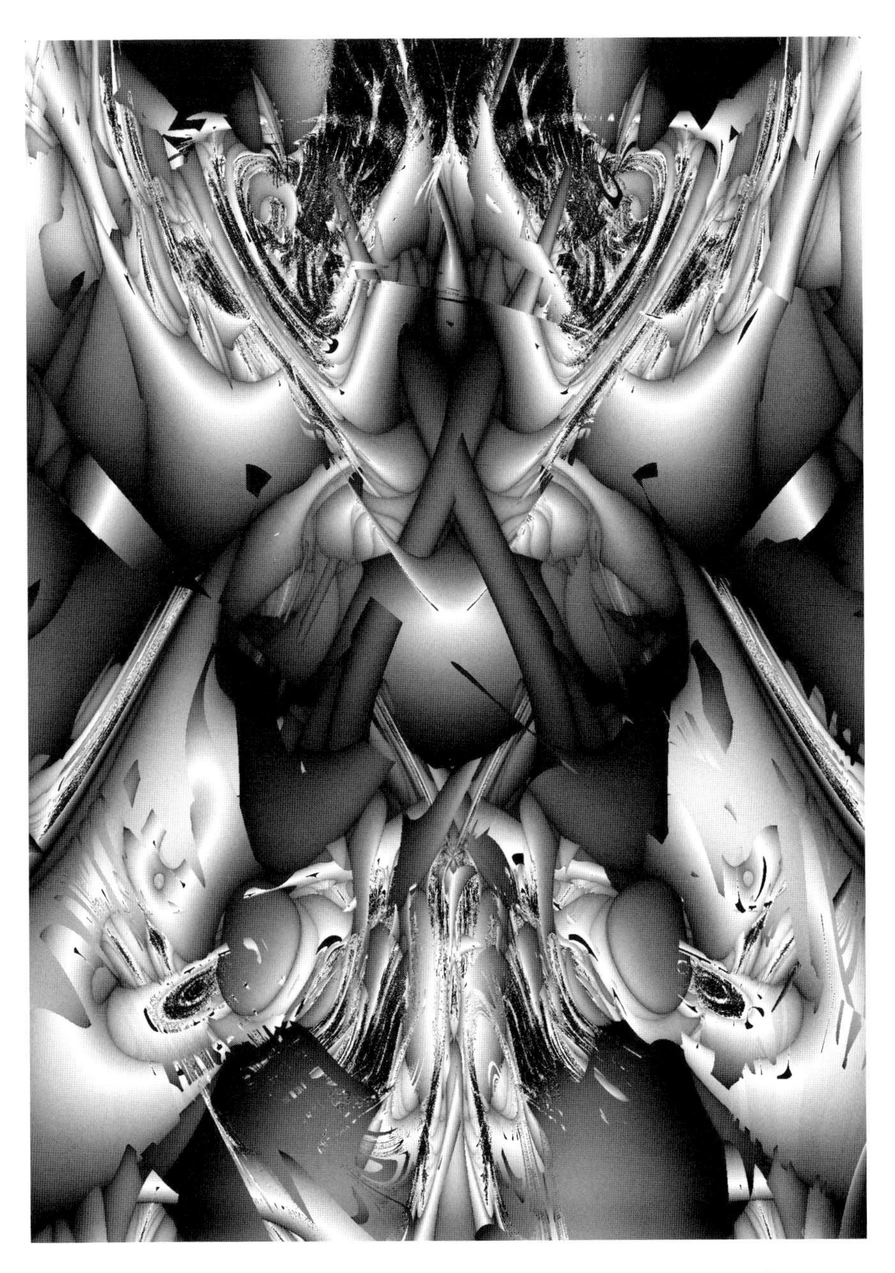

Fig. 9.40 Equation (9.48): $b = 1.6$, $k = 2$, $m = n = \gamma = \mu = 1$, $\alpha = 0$, $\beta = 1.5$, B versus A, r:AABB AABB..., $n_{\text{prev}} = 100$, $n_{\max} = 200$, $x_0 = 0.3$. D-shading. LL:(3.793, 3.907), UL:(3.941, 4.055), LR:(3.905, 3.795)

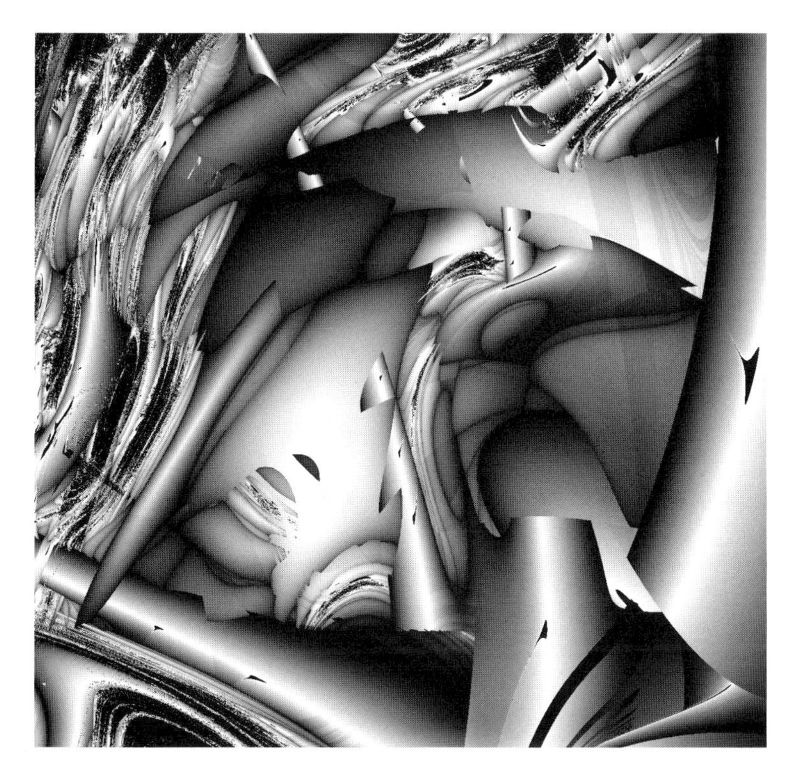

Fig. 9.41 Equation (9.48): $b = 1.8$, $k = 2$, $m = n = \gamma = \mu = 1$, $\alpha = 0$, $\beta = 1.5$, B versus A, r:AABB AABB..., $n_{\text{prev}} = 200$, $n_{\max} = 200$, $x_0 = 0.5$. D-shading. LL:(2.305, 2.357), UL:(2.305, 2.415), LR:(2.365, 2.357)

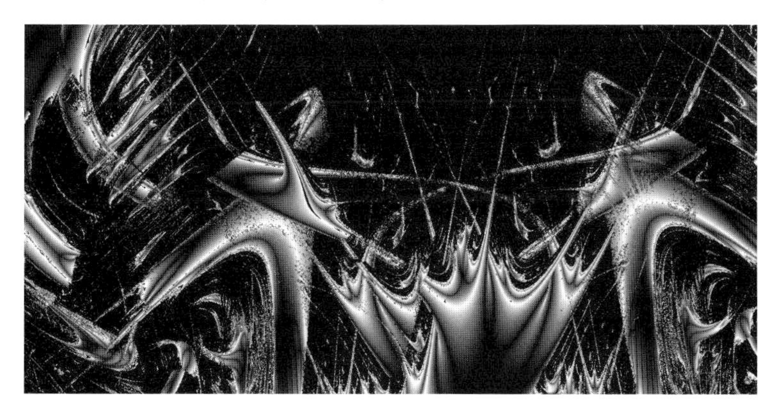

Fig. 9.42 Equation (9.48): $b = 2.3$, $\gamma = \mu = k = m = 1$, $n \to -\infty$, $\alpha = 0.2$, $\beta = 0$, B versus A, r:AAABBB AAABBB..., $n_{\text{prev}} = 100$, $n_{\max} = 200$, $x_0 = 0.5$. D-shading. LL:(0.13, −1.0137), UL:(−0.3237, −1.4212), LR:(−0.7637, −0.01868)

Fig. 9.43 Equation (9.48): $b = 1.8$, $k = 2$, $m = n = \gamma = \mu = 1$, $\alpha = 0$, $\beta = 1.5$, B versus A, r:AAABBB AAABBB..., $n_{\text{prev}} = 1000$, $n_{\max} = 1000$, $x_0 = 0.5$. D-shading. LL:(2.091, 2.3), UL:(2.349, 2.558), LR:(2.3, 2.091)

Fig. 9.44 Equation (9.48): $b = 2$, $k = 2$, $m = n = \gamma = \mu = 1$, $\alpha = 0$, $\beta = 1.25$, B versus A, r:AAABBB AAABBB..., $n_{\mathrm{prev}} = 200$, $n_{\mathrm{max}} = 200$, $x_0 = 1.5$. D-shading. LL:(5.827, 5.879), UL:(5.864, 5.916), LR:(5.879, 5.827)

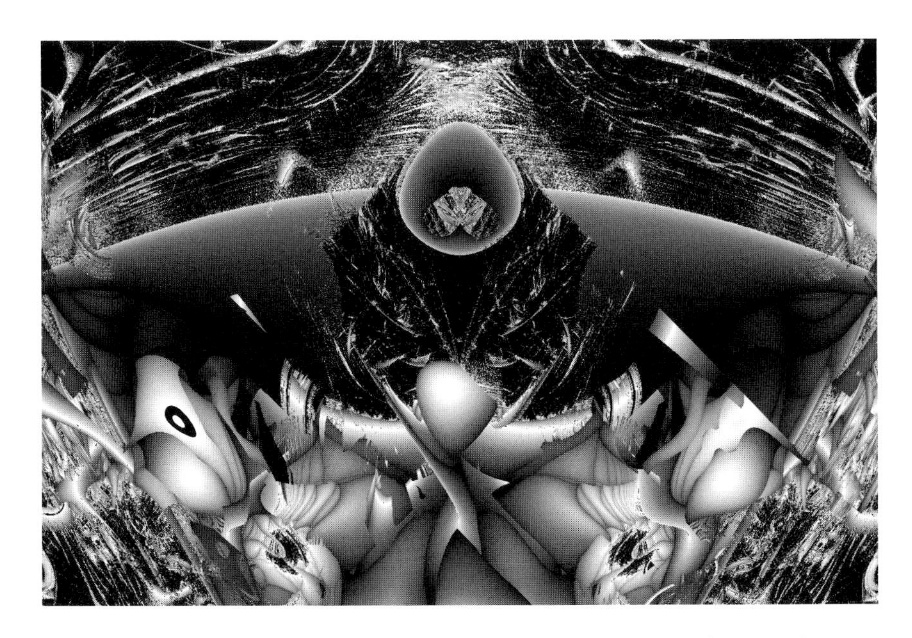

Fig. 9.45 Equation (9.48): $b = 2$, $k = 2$, $m = n = \gamma = \mu = 1$, $\alpha = 0$, $\beta = 1.25$, B versus A, r:AAABBB AAABBB..., $n_{\text{prev}} = 500$, $n_{\text{max}} = 500$, $x_0 = 1.5$. D-shading. LL:(3.002, 3.27), UL:(3.179, 3.449), LR:(3.272, 3.003)

Fig. 9.46 Equation (9.48): $b = 1.6$, $\gamma = \mu = k = m = 1$, $n \to -\infty$, $\alpha = 0$, $\beta = 1.5$, B versus A. r:AABB AABB..., $n_{\text{prev}} = 200$, $n_{\text{max}} = 200$, $x_0 = 0.3$. D-shading. LL:(2.7, 3.8925), UL:(3.1475, 4.34), LR:(3.8925, 2.7)

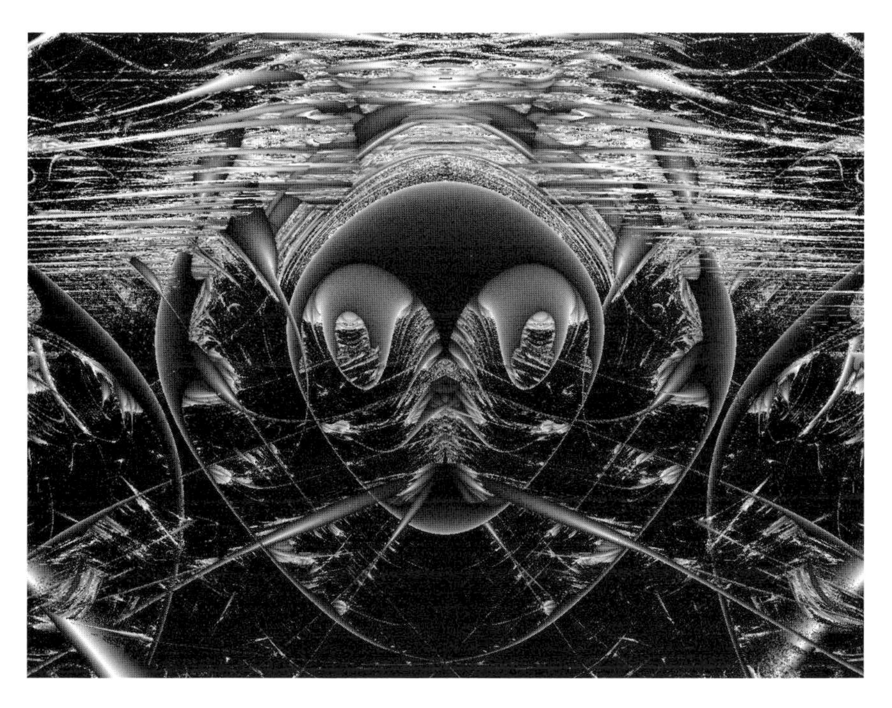

Fig. 9.47 Equation (9.48): $b = 2$, $k = 2$, $m = n = \gamma = \mu = 1$, $\alpha = -0$, $\beta = 1.25$, B versus A, r:AABB AABB..., $n_{\text{prev}} = 500$, $n_{\text{max}} = 500$, $x_0 = 1.5$. D-shading. LL:(3.203, 3.257), UL:(3.28, 3.334), LR:(3.257, 3.203)

Fig. 9.48 Equation (9.48): $b = 1.7$, $k = 2$, $m = n = \gamma = \mu = 1$, $\alpha = 0$, $\beta = 1.5$, B versus A, r:AABB AABB..., $n_{\text{prev}} = 200$, $n_{\max} = 200$, $x_0 = 0.3$. D-shading. LL:(5.6016, 5.7464), UL:(5.7962, 5.941), LR:(5.7464, 5.6016)

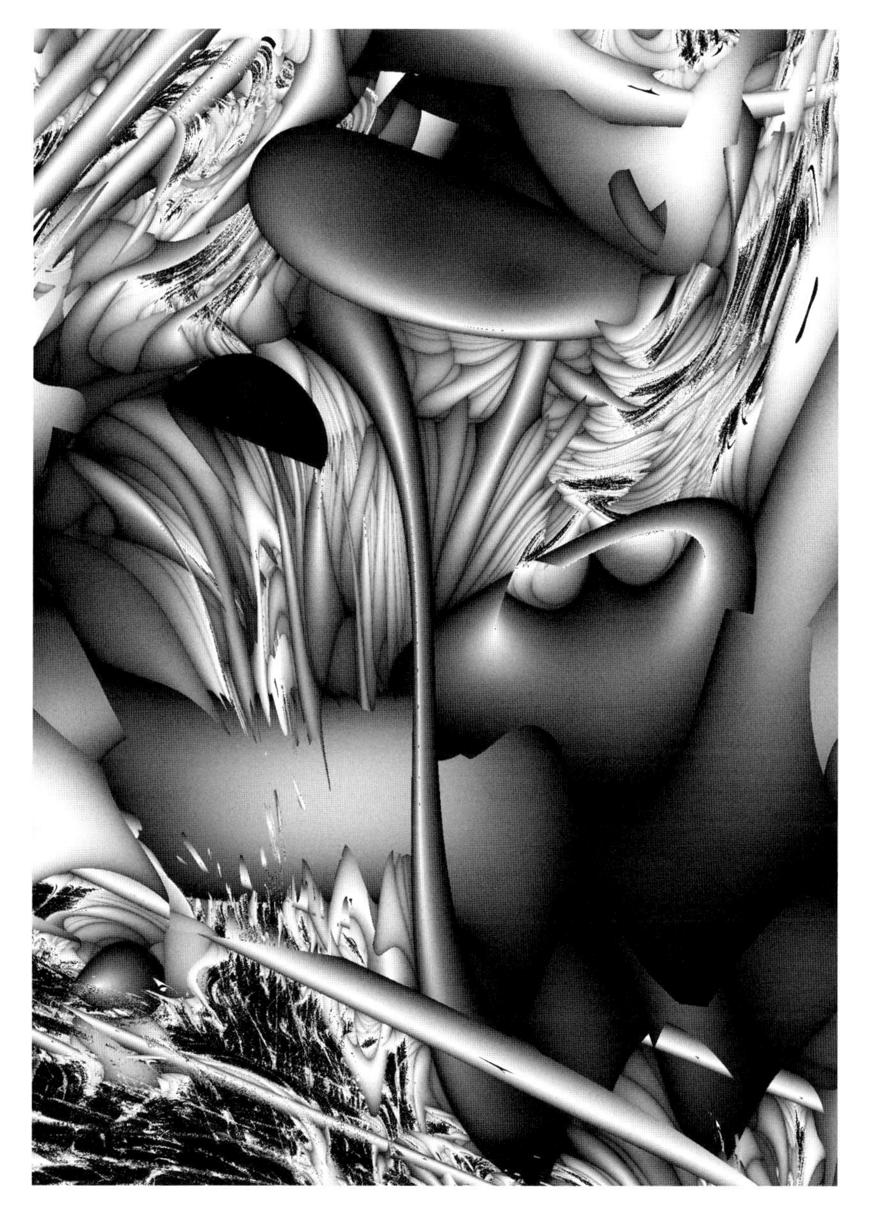

Fig. 9.49 Equation (9.48): $b = 1.7$, $k = 2$, $m = n = \gamma = \mu = 1$, $\alpha = 0$, $\beta = 1.5$, B versus A, r:AABB AABB..., $n_{\text{prev}} = 200$, $n_{\text{max}} = 200$, $x_0 = 0.3$. D-shading. LL:(3.394, 3.278), UL:(3.6105, 3.4185), LR:(3.4825, 3.1416)

Fig. 9.50 Equation (9.48): $b = 2.25$, $k = 2$, $m = n = \gamma = \mu = 1$, $\alpha = 1$, $\beta = 0$, B versus A, r:AABB AABB..., $n_{\text{prev}} = 200$, $n_{\text{max}} = 200$, $x_0 = 0.5$. D-shading. LL:(2.446, 1.074), UL:(2.446, 1.416), LR:(3.18, 1.074)

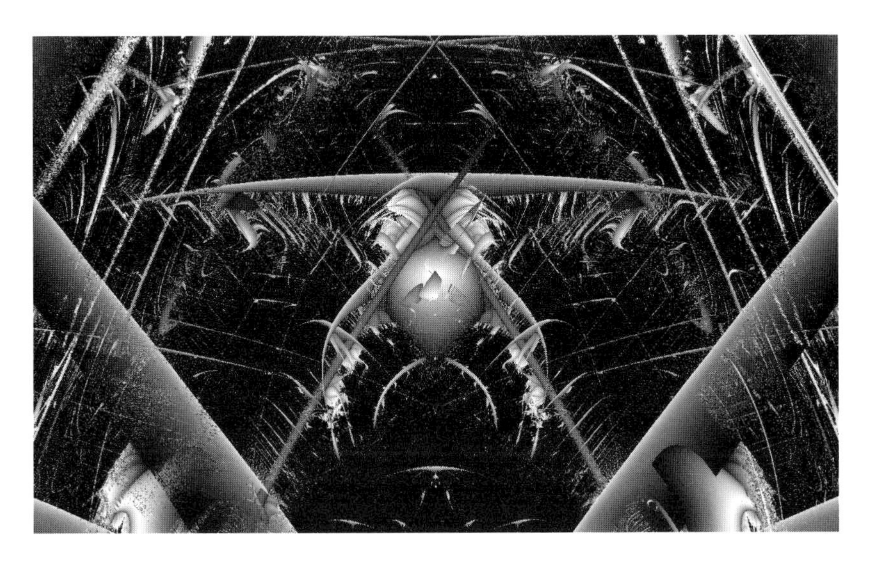

Fig. 9.51 Equation (9.48): $b = 2$, $k = m = n = \gamma = \mu = 1$, $\alpha = 0.7$, $\beta = 0$, B versus A, r:AABB AABB..., $n_{\text{prev}} = 200$, $n_{\text{max}} = 1000$, $x_0 = 0.5$. D-shading. LL:(0.95, 1.1425), UL:(1.0325, 1.225), LR:(1.1425, 0.95)

Fig. 9.52 Equation (9.48): $b = 2.3$, $m = n = \gamma = \mu = 1$, $\alpha = \beta = 0$, i.e. conditions of Eq...., B versus A, r:AAABBB AAABBB..., $n_{\text{prev}} = 300$, $n_{\max} = 400$, $x_0 = 0.3$. D-shading. LL:(−0.2, 1.47325), UL:(1.47325, −0.2), LR:(0.93675, 2.61)

Fig. 9.53 Equation (9.48): $b = 2$, $k = m = n = \gamma = \mu = 1$, $\alpha = -0.4$, $\beta = 1.5$, B versus A, r:AABB AABB..., $n_{\text{prev}} = 600$, $n_{\max} = 600$, $x_0 = 0.5$. D-shading. LL:(5.15, 5.405), UL:(5.414, 5.669), LR:(5.405, 5.15)

Fig. 9.54 Equation (9.48): $b = 2$, $k = m = n = \gamma = \mu = 1$, $\alpha = 0.7$, $\beta = 1.25$, B versus A, r:AABB AABB..., $n_{\text{prev}} = 800$, $n_{\text{max}} = 800$, $x_0 = 0.5$. D-shading. LL:(1.395, 1.863), UL:(1.771, 2.24), LR:(1.863, 1.3962)

Fig. 9.55 Equation (9.48): $b = 2.3$, $k = m = n = \gamma = \mu = 1$, $\alpha = 0$, $\beta = 0.25$, B versus A, r:AAABBB AAABBB..., $n_{\rm prev} = 400$, $n_{\max} = 1000$, $x_0 = 0.3$. D-shading. LL:(−0.2, 1.2), UL:(0.8, 2.2), LR:(1.2, −0.2)

Fig. 9.56 Equation (9.48): $b = 2.3$, $k = m = n = \gamma = \mu = 1$, $\alpha = 0$, $\beta = 0.15$, B versus A, r:AAABBB AAABBB..., $n_{\text{prev}} = 300$, $n_{\text{max}} = 300$, $x_0 = 0.5$. D-shading. LL:$(-1.52, -0.93)$, UL:$(-2.2, -0.93)$, LR:$(-1.52, -0.25)$

Fig. 9.57 Equation (9.48): $b = 2.3$, $k = m = n = \gamma = \mu = 1$, $\alpha = 0$, $\beta = 0.25$, B versus A, r:AAABBB AAABBB..., $n_{\text{prev}} = 200$, $n_{\text{max}} = 500$, $x_0 = 0.3$. D-shading. LL:(4.22, 6.775), UL:(5.285, 6.4589), LR:(3.802, 5.3667)

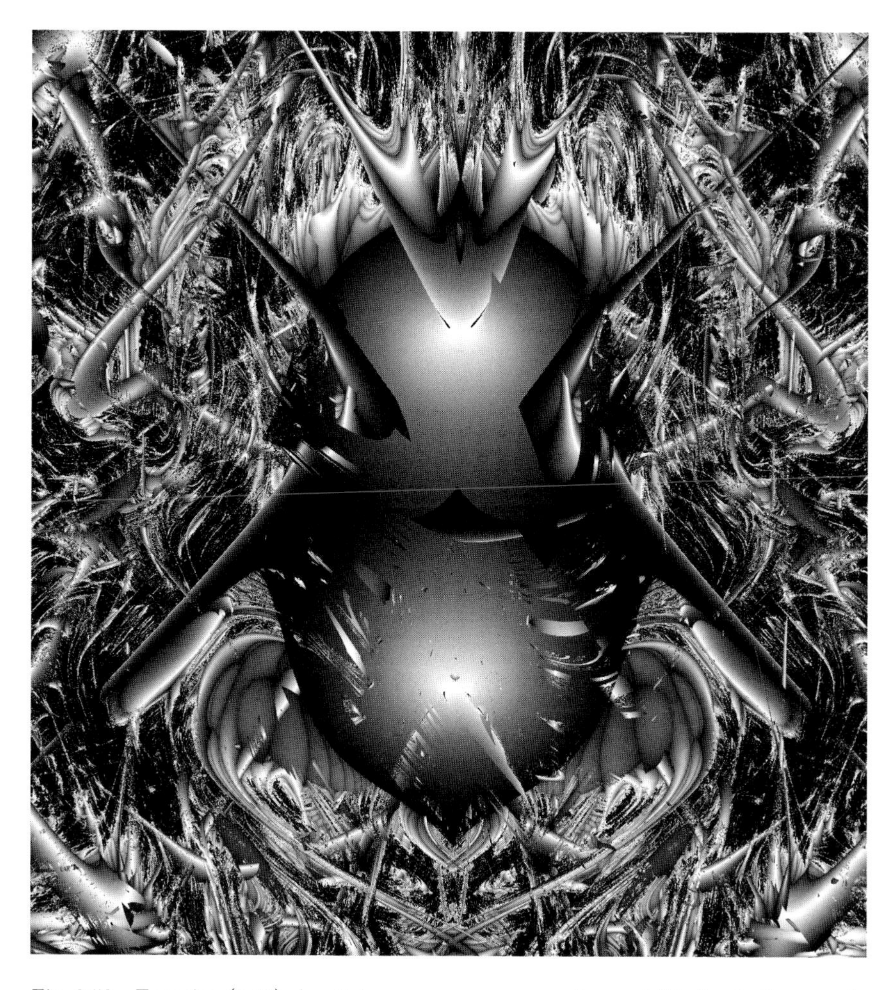

Fig. 9.58 Equation (9.48): $b = 2$, $k = m = n = \gamma = \mu = 1$, $\alpha = 1.25$, $\beta = 0$, B versus A, r:AABB AABB..., $n_{\text{prev}} = 100$, $n_{\max} = 200$, $x_0 = 1.5$. D-shading. LL:$(-1.26, -0.08)$, UL:$(-0.339, 0.861)$, LR:$(-0.08, -1.2349)$

Fig. 9.59 Equation (9.48): $b = 1.8$, $k = m = n = \gamma = \mu = 1$, $\alpha = 0$, $\beta = 1.5$, B versus A, r:AABB AABB..., $n_{\text{prev}} = 100$, $n_{\max} = 200$, $x_0 = 0.5$. D-shading. LL:(3.95, 6.1), UL:(3.95, 7.8), LR:(4.87, 6.1)

Fig. 9.60 Equation (9.48): $b = 1.7$, $k = m = n = \gamma = \mu = 1$, $\alpha = 0$, $\beta = 1.5$, B versus A, r:AABB AABB..., $n_{\text{prev}} = 100$, $n_{\max} = 200$, $x_0 = 0.3$. D-shading. LL:$(-1.68, -0.64)$, UL:$(-0.64, 0.4)$, LR:$(-0.64, -1.68)$

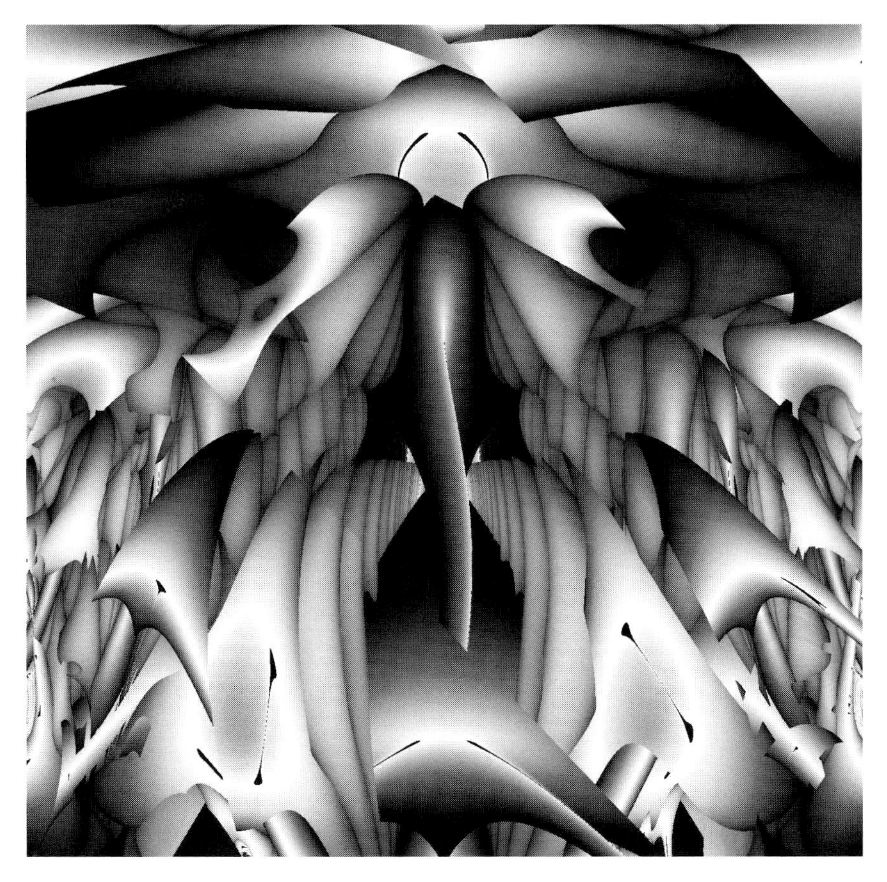

Fig. 9.61 Equation (9.48): $b = 1.6$, $k = m = n = \gamma = \mu = 1$, $\alpha = 0$, $\beta = 1.5$, B versus A, r:AABB AABB..., $n_{\text{prev}} = 100$, $n_{\text{max}} = 200$, $x_0 = 0.3$. D-shading. LL:(1.885, 2.55), UL:(2.52, 3.185), LR:(2.55, 1.885)

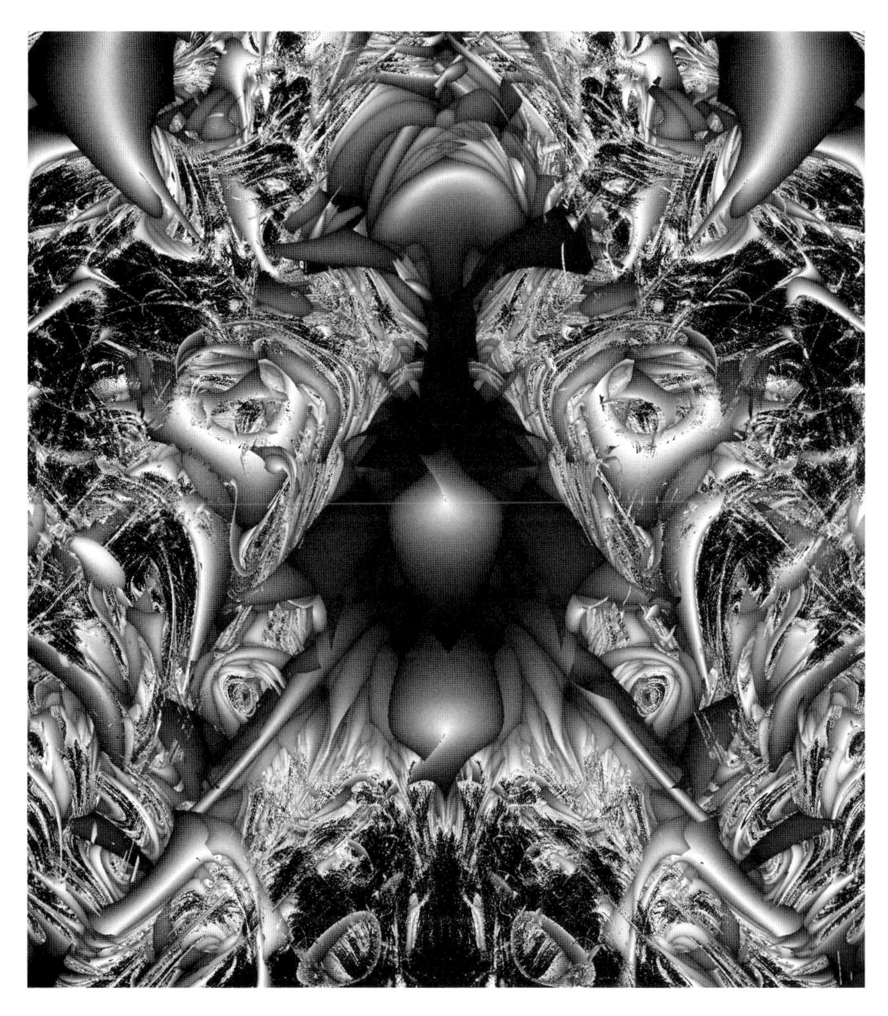

Fig. 9.62 Equation (9.48): $b = 1.8$, $k = m = n = \gamma = \mu = 1$, $\alpha = 0$, $\beta = 1.5$, B versus A, r:AAABBB AAABBB..., $n_{\text{prev}} = 100$, $n_{\max} = 200$, $x_0 = 0.5$. D-shading. LL:(0.16, 1.41), UL:(1.26, 2.51), LR:(1.41, 0.16)

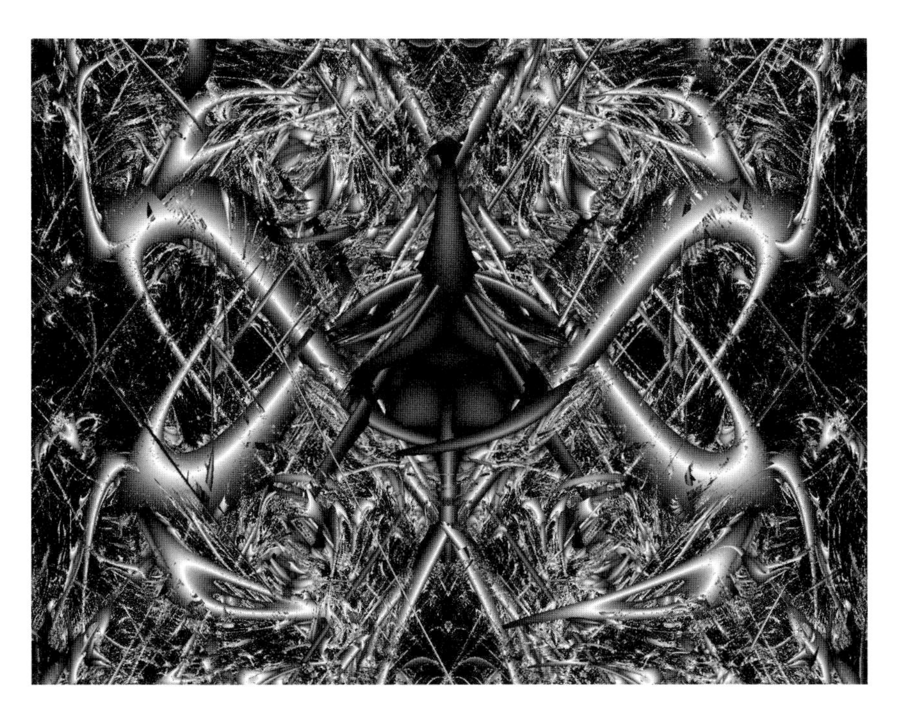

Fig. 9.63 Equation (9.48): $b = 2$, $k = m = n = \gamma = \mu = 1$, $\alpha = 2.4$, $\beta = 0$, B versus A, r:AABB AABB..., $n_{\mathrm{prev}} = 200$, $n_{\mathrm{max}} = 400$, $x_0 = 0.5$. D-shading. LL:(0.37, 1.67), UL:(0.97, 2.27), LR:(1.67, 0.37)

Fig. 9.64 Equation (9.48): $b = 2$, $k = m = n = \gamma = \mu = 1$, $\alpha = 0$, $\beta = 1.25$, B versus A, r:AABB AABB..., $n_{\text{prev}} = 100$, $n_{\max} = 200$, $x_0 = 1.5$. D-shading. LL:(2.29, 3.19), UL:(2.887, 3.787), LR:(3.19, 2.29)

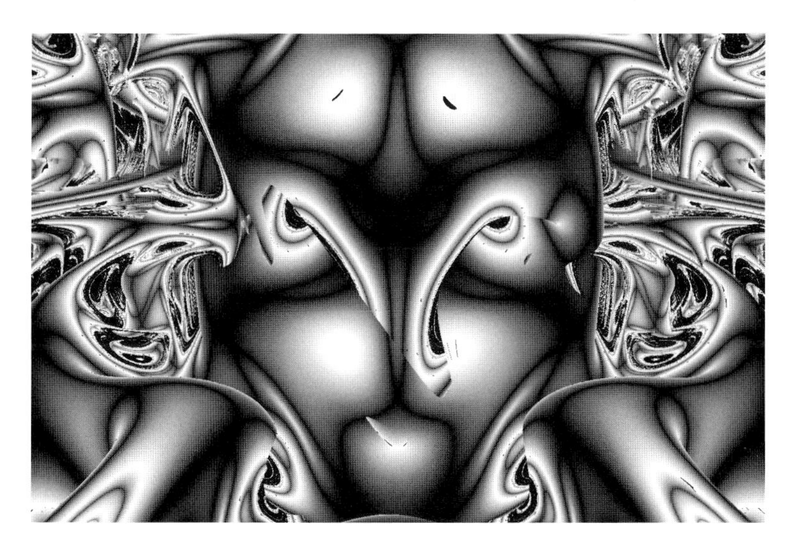

Fig. 9.65 Equation (9.48): $b = 1.6$, $\gamma = \mu = k = m = 1$, $n \to -\infty$, $\alpha = 2.8$, $\beta = 0$, B versus A, r:AABB AABB..., $n_{\text{prev}} = 100$, $n_{\max} = 200$, $x_0 = 0.39$. D-shading. LL:(-1.37, -0.22), UL:(-0.75, 0.4), LR:(-0.22, -1.37)

Fig. 9.66 Equation (9.48): $b = 2.25$, $k = m = n = \gamma = \mu = 1$, $\alpha = 0$, $\beta = -1$, B versus A, r:AABB AABB..., $n_{\text{prev}} = 100$, $n_{\text{max}} = 200$, $x_0 = 0.5$. D-shading. LL:(4.06, 4.151), UL:(4.143, 4.234), LR:(4.151, 4.06)

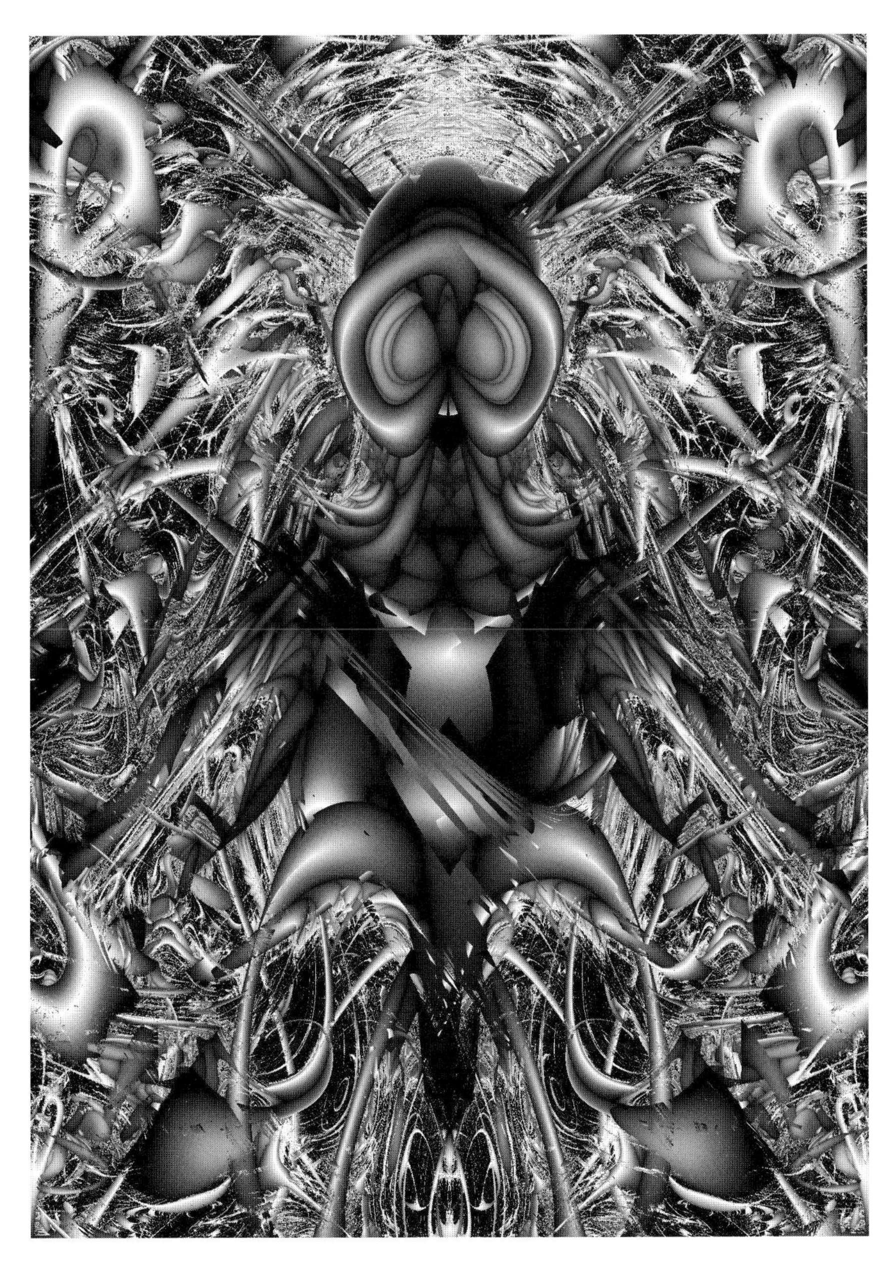

Fig. 9.67 Equation (9.48): $b = 2$, $k = m = n = \gamma = \mu = 1$, $\alpha = 0$, $\beta = 1.25$, B versus A, r:AAABBB AAABBB..., $n_{\text{prev}} = 1000$, $n_{\max} = 1000$, $x_0 = 1.5$. D-shading. LL:(1.6, 2.76), UL:(2.83, 3.99), LR:(2.76, 1.6)

Fig. 9.68 Equation (9.48): $b = 2.3$, $k = m = n = \gamma = \mu = 1$, $\alpha = 0.1$, $\beta = 0$, B versus A, r:AABB AABB..., $n_{\text{prev}} = 100$, $n_{\max} = 200$, $x_0 = 0.7$. D-shading. LL:(1.63, 2.655), UL:(2.655, 3.68), LR:(2.705, 1.58)

Fig. 9.69 Equation (9.48): $b = 1.5$, $k = m = n = \gamma = \mu = 1$, $\alpha = -0.4$, $\beta = 0$, B versus A, r:AABB AABB..., $n_{\text{prev}} = 200$, $n_{\text{max}} = 400$, $x_0 = 0.5$. D-shading. LL:(1.19, 1.87), UL:(2.19, 2.96), LR:(1.95, 1.1728)

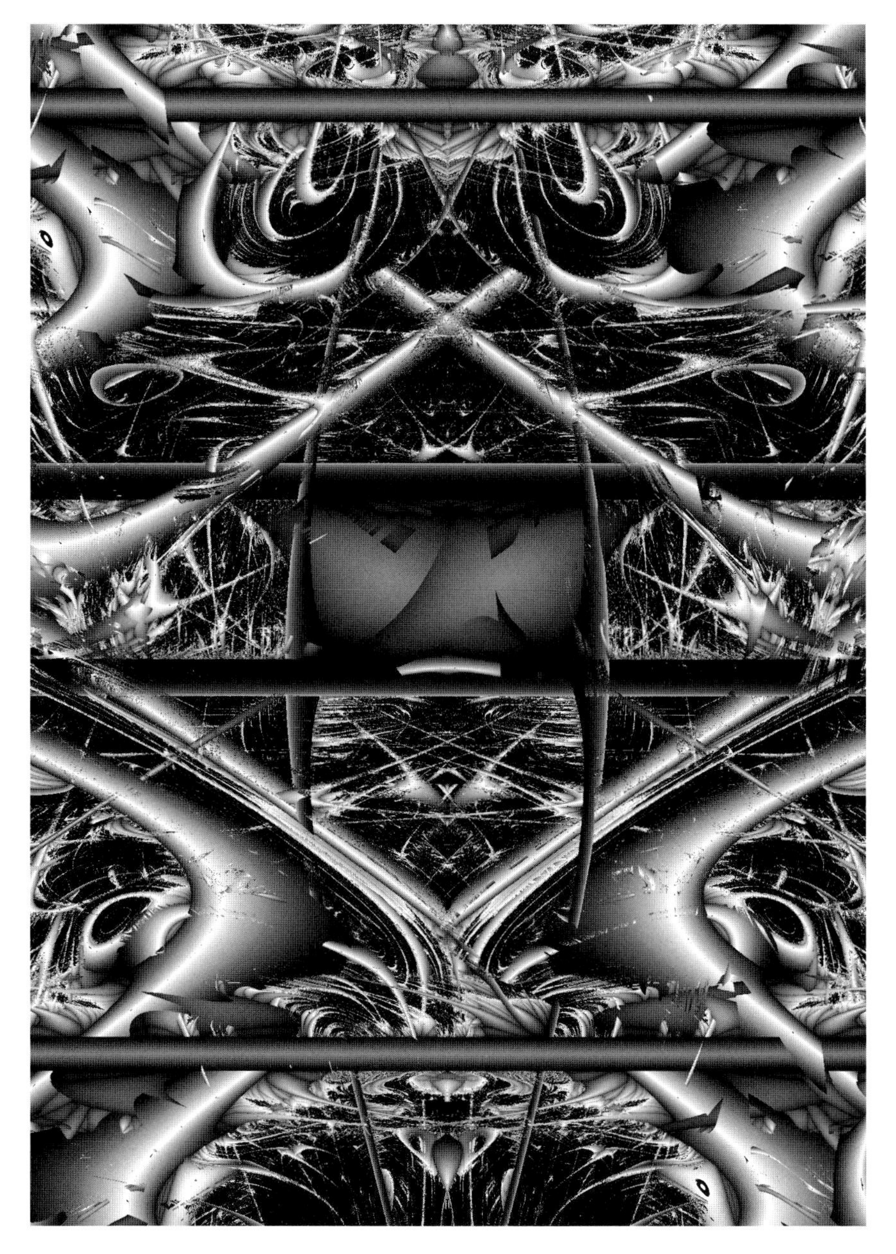

Fig. 9.70 Equation (9.48): $b = 2.1$, $k = m = n = \gamma = \mu = 1$, $\alpha = 1$, $\beta = 0$, B versus A, r:AB AB..., $n_{\text{prev}} = 200$, $n_{\max} = 200$, $x_0 = 0.4$. D-shading. LL:(2.51, 4.51), UL:(4.5, 6.5), LR:(4.51, 2.51)

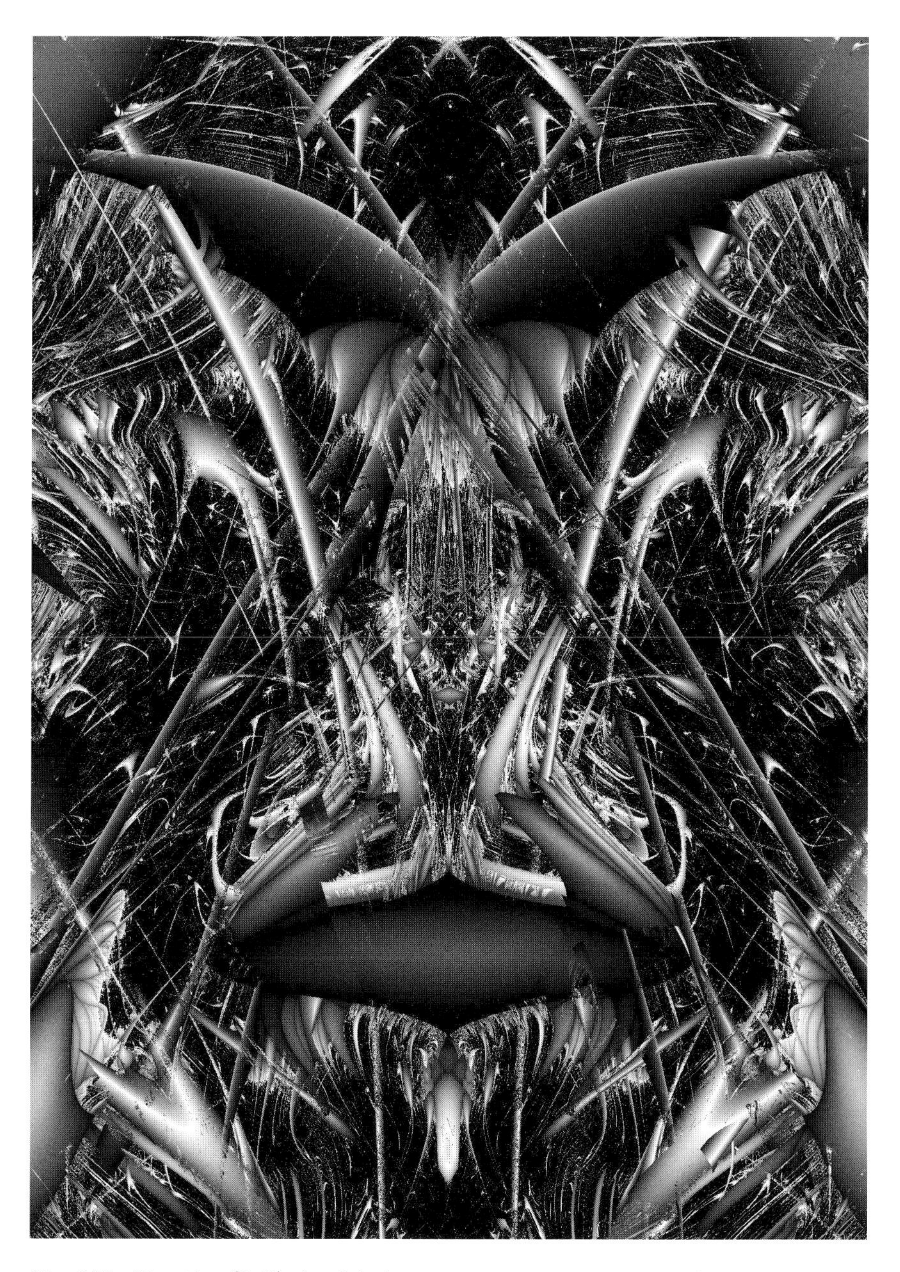

Fig. 9.71 Equation (9.48): $b = 2.1$, $k = m = n = \gamma = \mu = 1$, $\alpha = 1$, $\beta = 0$, B versus A, r:AABB AABB..., $n_{\text{prev}} = 1000$, $n_{\max} = 1000$, $x_0 = 0.4$. D-shading. LL:(0.333, 0.778), UL:(0.744, 1.189), LR:(0.778, 0.333)

Fig. 9.72 Equation (9.48): $b = 1$, $k = m = n = \gamma = \mu = 1$, $\alpha = 1$, $\beta = 0$, B versus A, r:AB AB..., $n_{\text{prev}} = 200$, $n_{\max} = 200$, $x_0 = 0.4$. D-shading. LL:(0, 0.05), UL:(0, 9.95), LR:(10, 0.05)

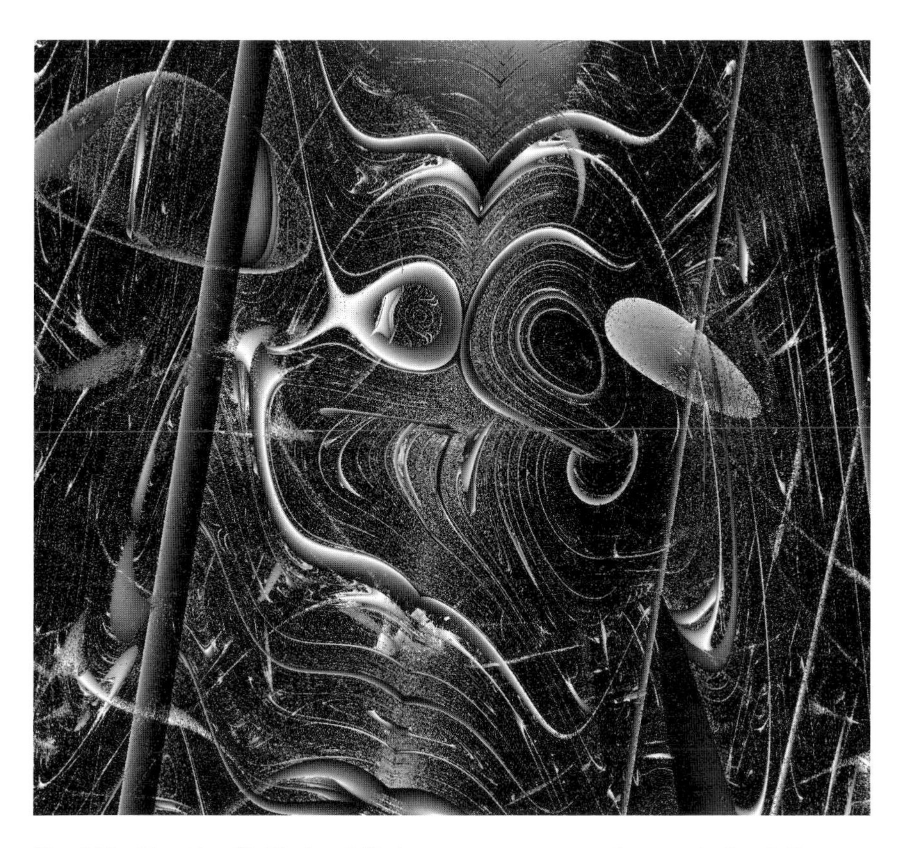

Fig. 9.73 Equation (9.48): $b = 2.75$, $k = m = n = \gamma = \mu = 1$, $\alpha = -1$, $\beta = 0$, B versus A, r:AAABB AAABB..., $n_{\text{prev}} = 200$, $n_{\max} = 400$, $x_0 = 0.6$. D-shading. LL:(0.702, 1.282), UL:(1.73, 2.51), LR:(1.451, 0.65499)

Fig. 9.74 Equation (9.48): $b = 2.25$, $k = m = n = \gamma = \mu = 1$, $\alpha = -1$, $\beta = 0$, B versus A, r:AABB AABB..., $n_{\text{prev}} = 100$, $n_{\text{max}} = 400$, $x_0 = 0.5$. D-shading. LL:(0.195, 1.205), UL:(1.945, 2.955), LR:(1.205, 0.195)

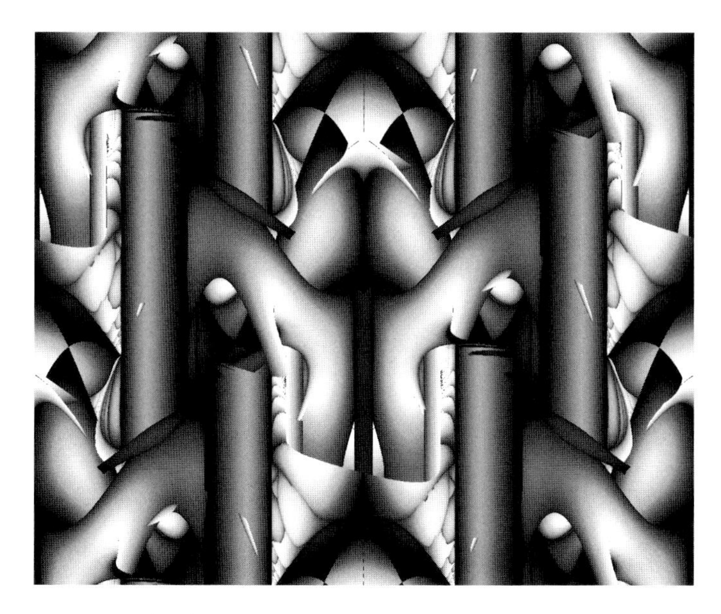

Fig. 9.75 Equation (9.48): $b = 1.5$, $k = m = n = \gamma = \mu = 1$, $\alpha = -1$, $\beta = 0$, B versus A, r:AABB AABB..., $n_{\text{prev}} = 100$, $n_{\max} = 200$, $x_0 = 0.5$. D-shading. LL:(1.2, 4.3), UL:(4.9, 8.), LR:(4.3, 1.2)

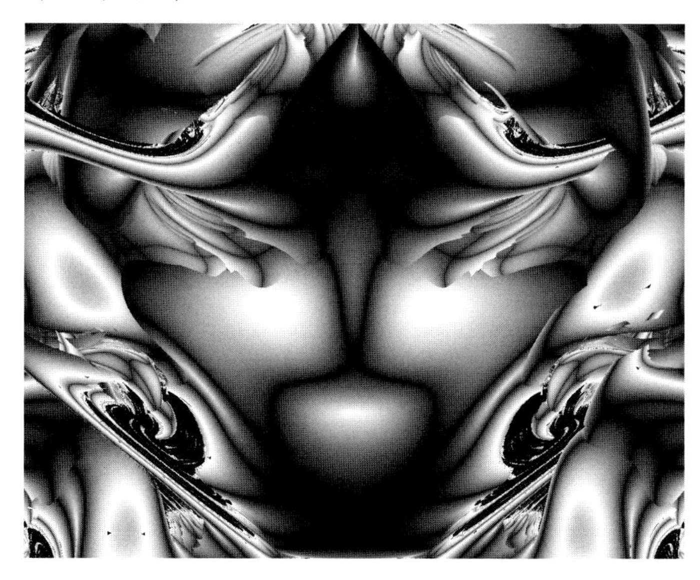

Fig. 9.76 Equation (9.48): $b = 1.6$, $k = m = n = \gamma = \mu = 1$, $\alpha = 2.8$, $\beta = 0$, B versus A, r:AABB AABB..., $n_{\text{prev}} = 100$, $n_{\max} = 200$, $x_0 = 0.39$. D-shading. LL:(3.729, 4.444), UL:(4.367, 5.082), LR:(4.444, 3.792)

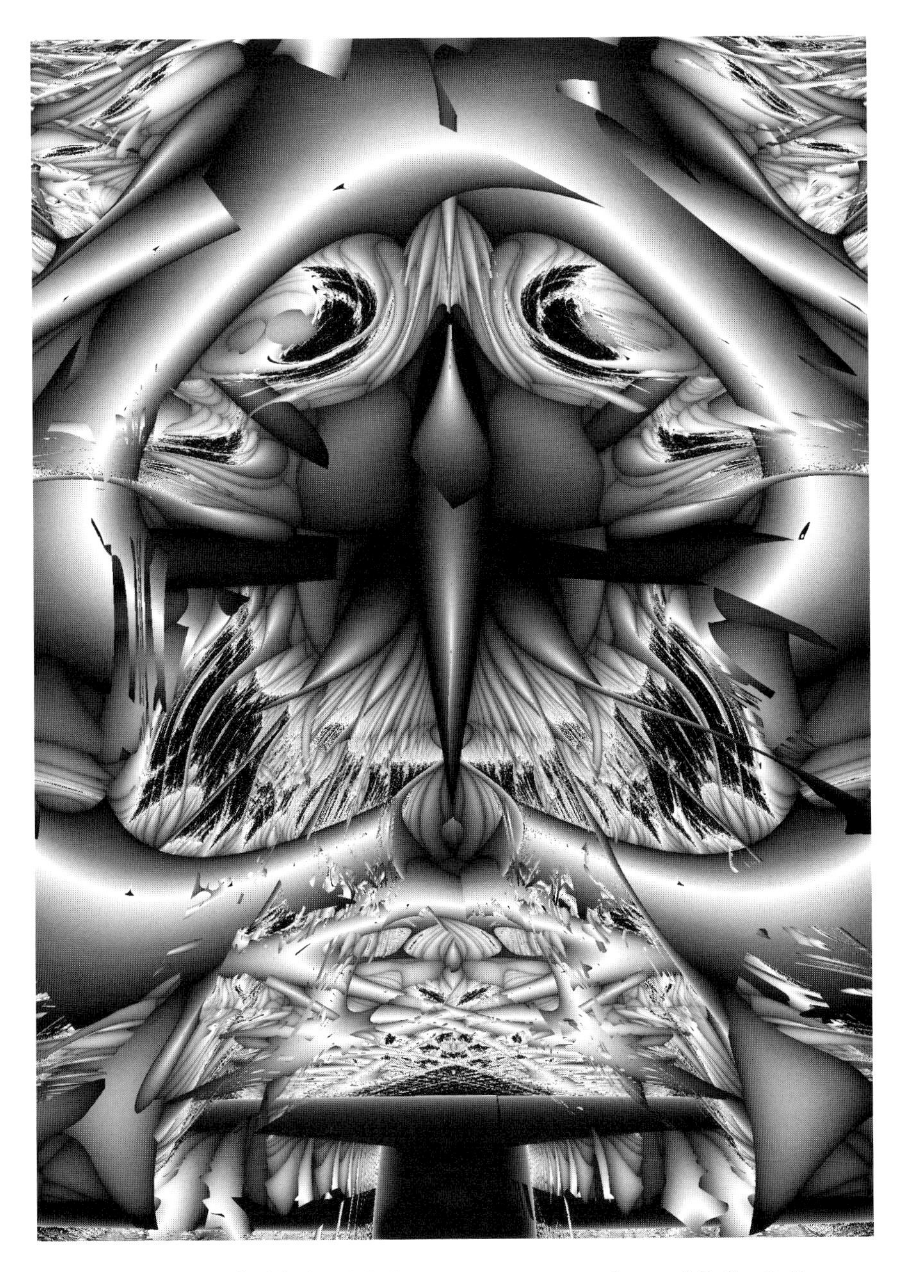

Fig. 9.77 Equation (9.48): $b = 1.6$, $k = m = n = \gamma = \mu = 1$, $\alpha = 0.7$, $\beta = 0$, B versus A, r:AB AB..., $n_{\text{prev}} = 100$, $n_{\max} = 200$, $x_0 = 0.3$. D-shading. LL:(0.87, 2.58), UL:(2.7, 4.41), LR:(2.58, 0.87)

Fig. 9.78 Equation (9.48): $b = 2$, $k = m = n = \gamma = \mu = 1$, $\alpha = -0.4$, $\beta = 0$, B versus A, r:AB AB..., $n_{\text{prev}} = 2000$, $n_{\max} = 2000$, $x_0 = 0.3$. D-shading. LL:(3.606, 3.954), UL:(3.981, 4.329), LR:(3.954, 3.606)

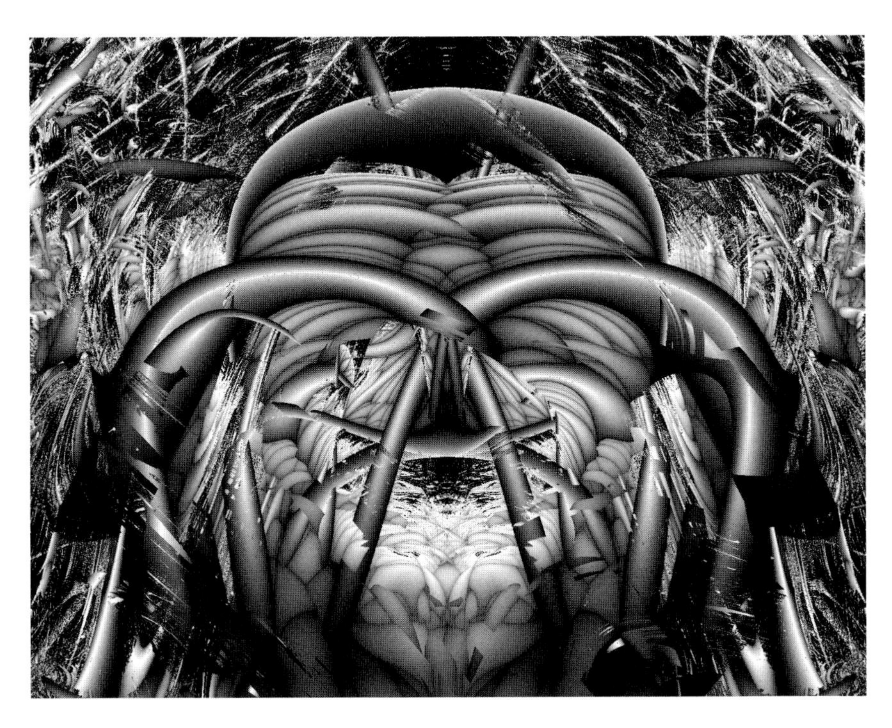

Fig. 9.79 Equation (9.48): $b = 2$, $k = m = n = \gamma = \mu = 1$, $\alpha = -0.2$, $\beta = 0$, B versus A, r:AAABBB AAABBB..., $n_{\text{prev}} = 400$, $n_{\text{max}} = 1000$, $x_0 = 0.5$. D-shading. LL:(3.23, 4.07), UL:(3.78, 4.62), LR:(4.07, 3.23)

Fig. 9.80 Equation (9.48): $b = 2$, $k = m = n = \gamma = \mu = 1$, $\alpha = -0.2$, $\beta = 0$, B versus A, r:AB AB..., $n_{\text{prev}} = 500$, $n_{\max} = 1500$, $x_0 = 0.5$. D-shading. LL:(4.34, 4.67), UL:(4.651, 4.981), LR:(4.67, 4.34)

Fig. 9.81 Equation (9.48): $b = 2$, $k = m = n = \gamma = \mu = 1$, $\alpha = -0.2$, $\beta = 0$, B versus A, r:AABB AABB..., $n_{\text{prev}} = 100$, $n_{\max} = 200$, $x_0 = 0.5$. D-shading. LL:(3.665, 3.955), UL:(4.045, 4.335), LR:(3.955, 3.665)

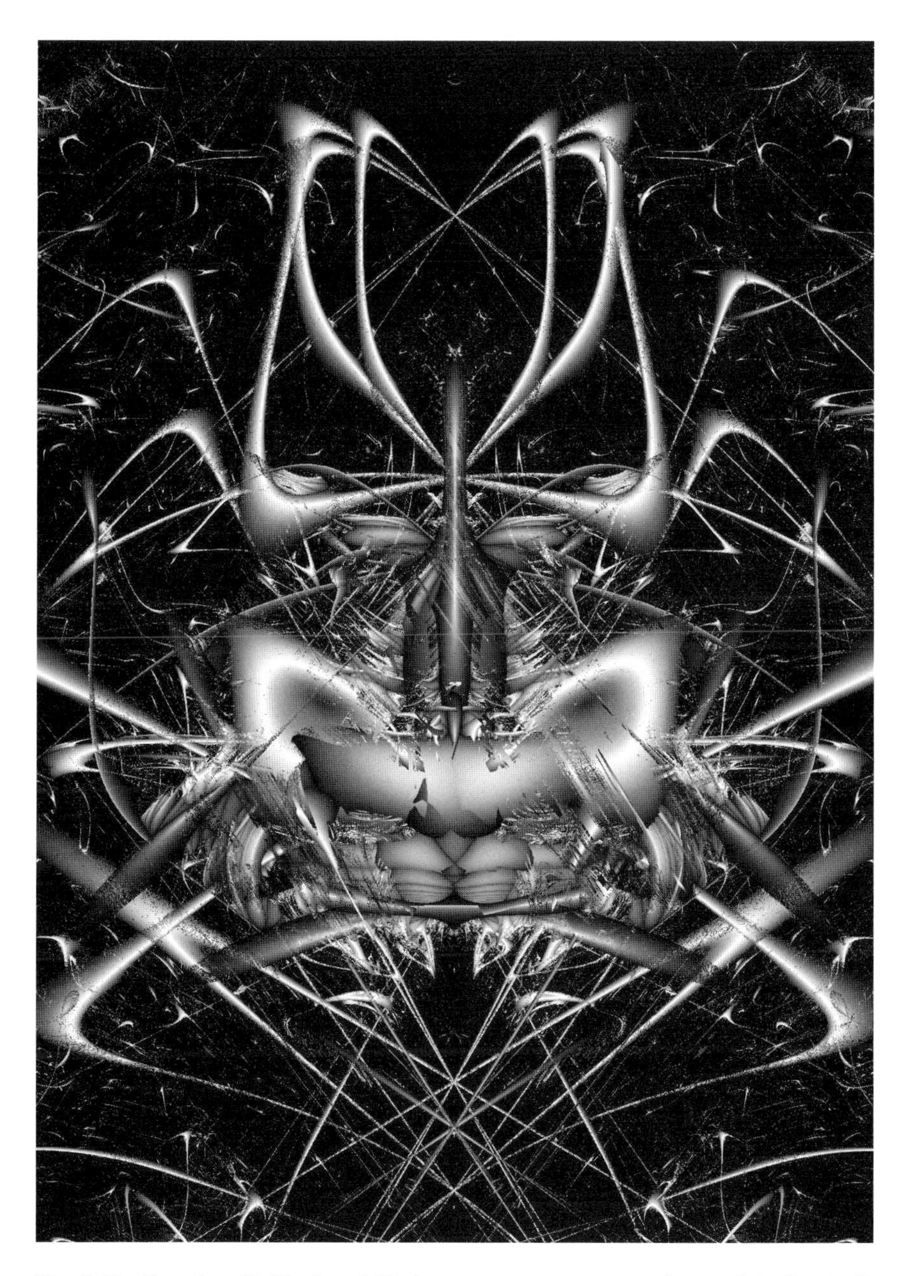

Fig. 9.82 Equation (9.48): $b = 2.75$, $k = m = n = \gamma = \mu = 1$, $\alpha = 0.2$, $\beta = 0$, B versus A, r:AAABBB AAABBB..., $n_{\text{prev}} = 1500$, $n_{\max} = 1500$, $x_0 = 0.5$. D-shading. LL:(0.265, 0.535), UL:(0.51, 0.78), LR:(0.535, 0.265)

Fig. 9.83 Equation (9.48): $b = 2.5$, $k = m = n = \gamma = \mu = 1$, $\alpha = 0.5$, $\beta = 0$, B versus A, r:AB AB..., $n_{\text{prev}} = 100$, $n_{\text{max}} = 200$, $x_0 = 0.5$. D-shading. LL:(1.95, 2.3915), UL:(2.6905, 3.144), LR:(2.398, 1.9506)

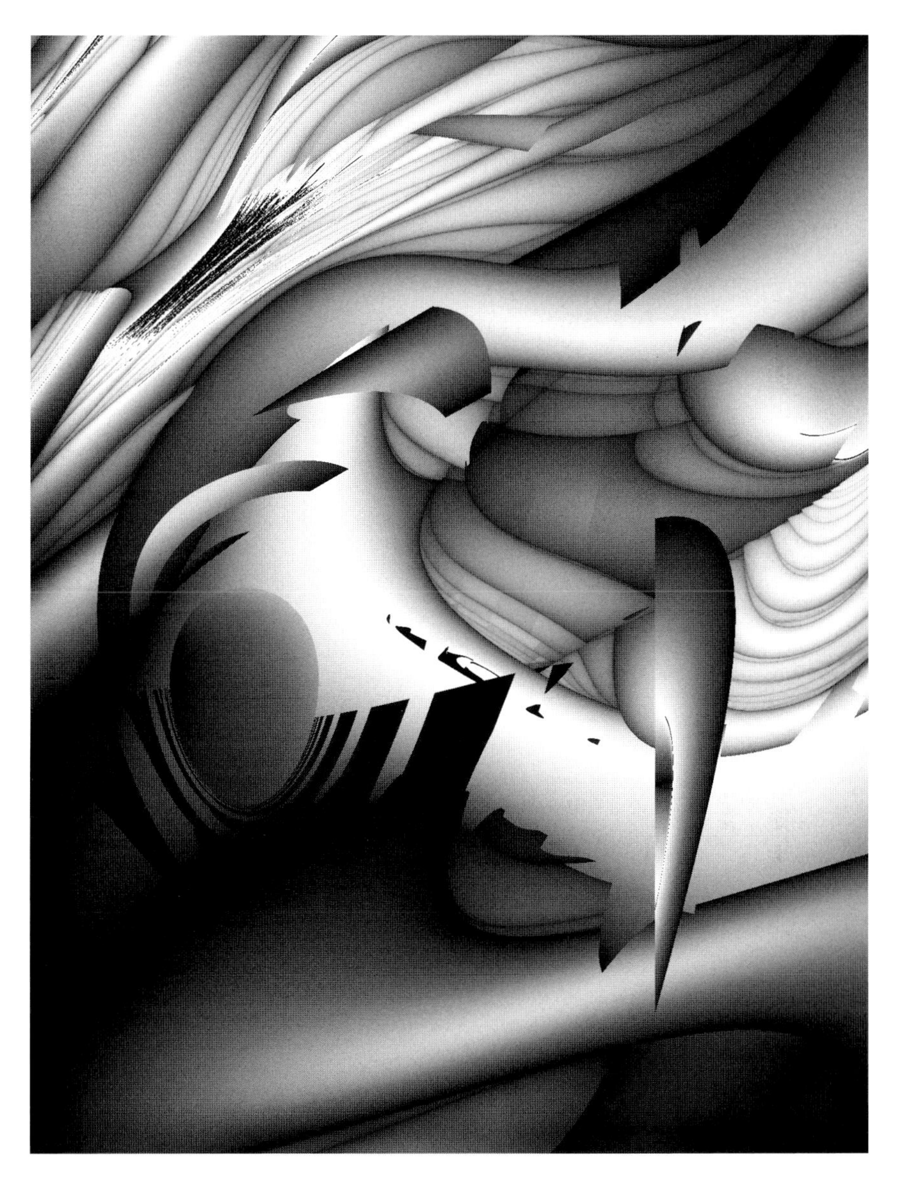

Fig. 9.84 Equation (9.48): $b = 1.5$, $k = m = n = \gamma = \mu = 1$, $\alpha = -1.25$, $\beta = 0$, B versus A, r:AABB AABB..., $n_{\text{prev}} = 100$, $n_{\text{max}} = 200$, $x_0 = 0.6$. D-shading. LL:(0.24, 4.92), UL:(0.24, 6.07), LR:(1.22, 4.92)

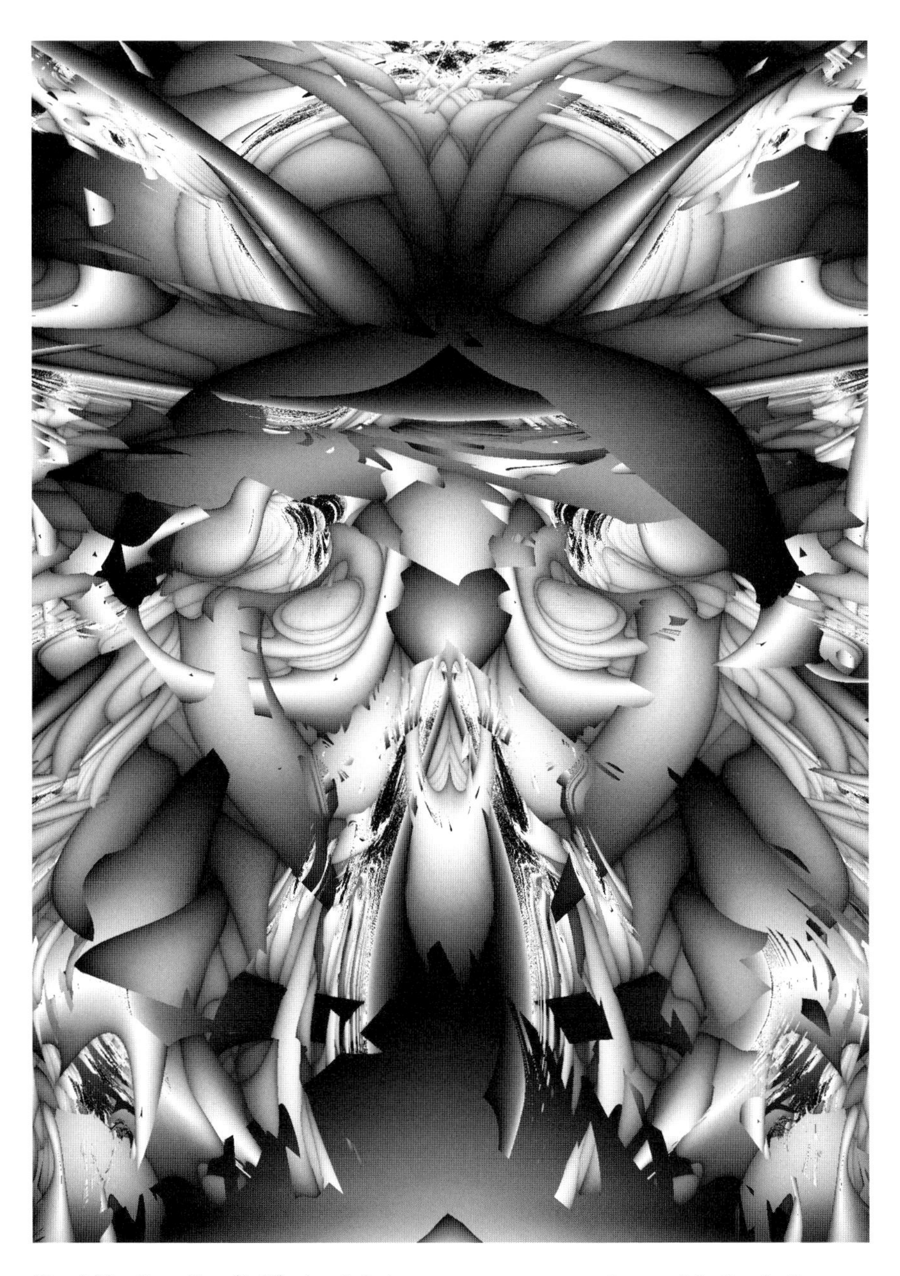

Fig. 9.85 Equation (9.48): $b = 1.5$, $k = m = n = \gamma = \mu = 1$, $\alpha = 0.6$, $\beta = 0$, B versus A, r:AABB AABB..., $n_{\text{prev}} = 100$, $n_{\text{max}} = 200$, $x_0 = 0.5$. D-shading. LL:(2.2, 3.2), UL:(3.9, 4.9), LR:(3.2, 2.2)

Fig. 9.86 Equation (9.48): $b = 2.5$, $k = m = n = \gamma = \mu = 1$, $\alpha = -2$, $\beta = 0$, B versus A, r:AABB AABB..., $n_{\text{prev}} = 400$, $n_{\max} = 1000$, $x_0 = 0.5$. D-shading. LL:(3.525, 3.885), UL:(3.824, 4.184), LR:(3.885, 3.525)

Fig. 9.87 Equation (9.48): $b = 2.3$, $k = m = n = \gamma = \mu = 1$, $\alpha = 0.25$, $\beta = 0$, B versus A, r:AAABBB AAABBB..., $n_{\text{prev}} = 200$, $n_{\max} = 400$, $x_0 = 0.3$. D-shading. LL:(5.46, 6.26), UL:(5.76, 6.56), LR:(6.26, 5.46)

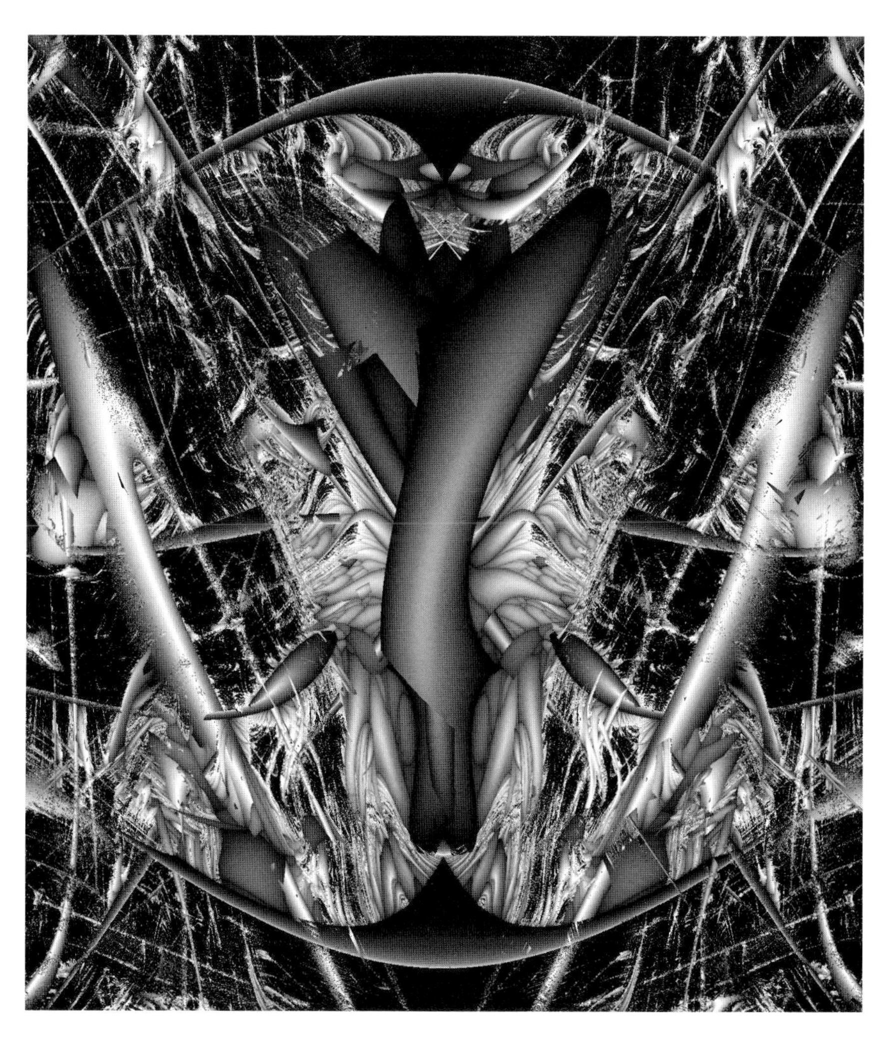

Fig. 9.88 Equation (9.48): $b = 2.25$, $k = m = n = \gamma = \mu = 1$, $\alpha = 1$, $\beta = 0$, B versus A, r:AABB AABB..., $n_{\text{prev}} = 100$, $n_{\max} = 200$, $x_0 = 0.5$. D-shading. LL:(0.778, 1.528), UL:(1.271, 2.021), LR:(1.528, 0.778)

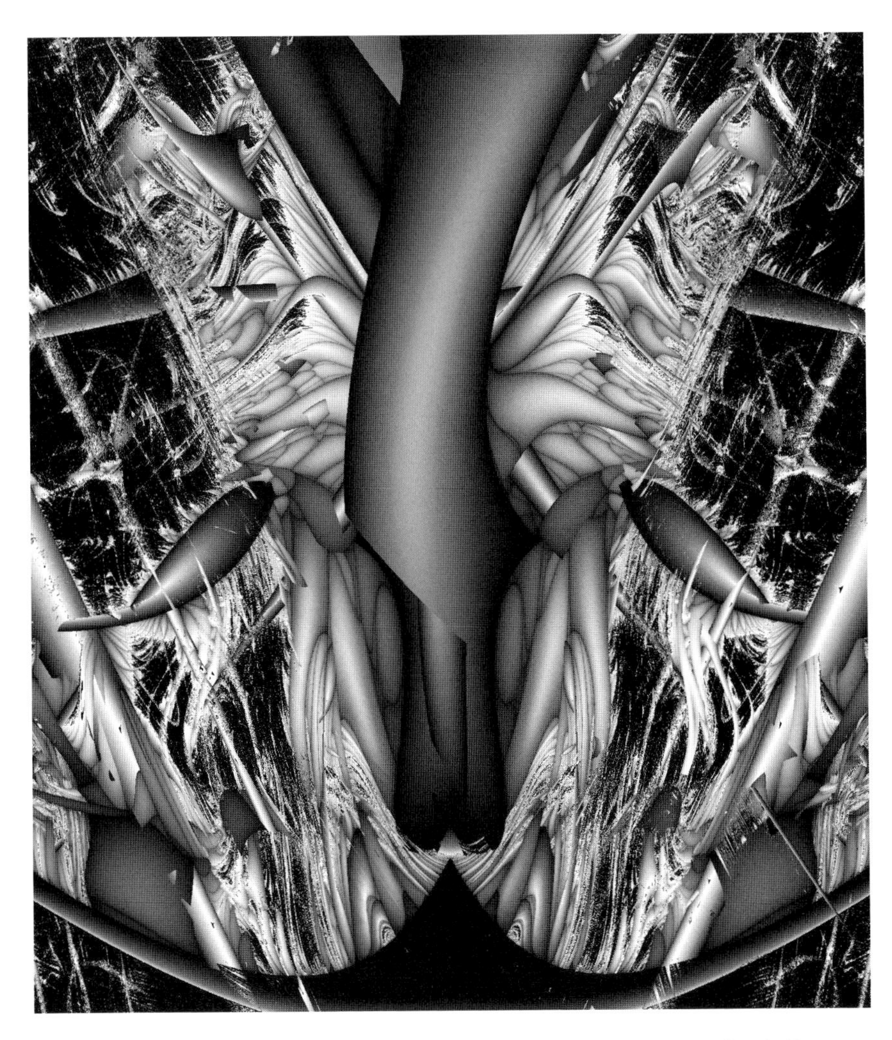

Fig. 9.89 Equation (9.48): $b = 2.25$, $k = m = n = \gamma = \mu = 1$, $\alpha = 1$, $\beta = 0$, B versus A, r:AABB AABB..., $n_{\mathrm{prev}} = 100$, $n_{\mathrm{max}} = 200$, $x_0 = 0.5$. D-shading. LL:(0.955, 1.415), UL:(1.249, 1.709), LR:(1.415, 0.955)

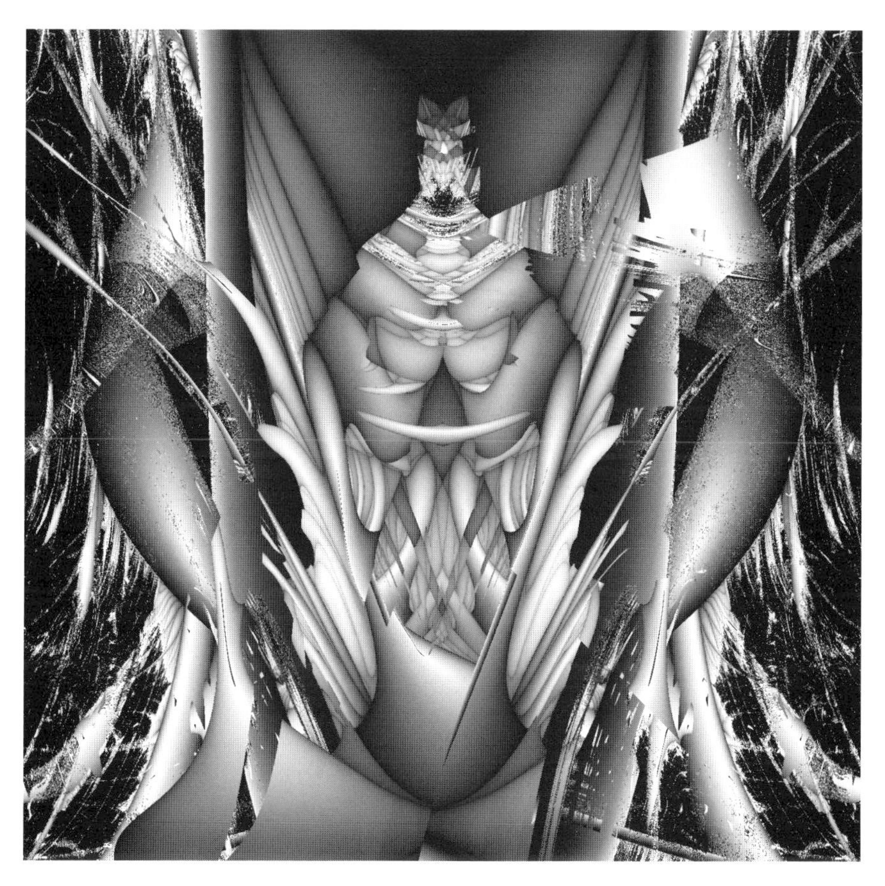

Fig. 9.90 Equation (9.48): $b = 2$, $k = m = n = \gamma = \mu = 1$, $\alpha = -0.4$, $\beta = 0$, B versus A, r:AABB AABB..., $n_{\text{prev}} = 100$, $n_{\max} = 200$, $x_0 = 0.3$. D-shading. LL:(1.6, 2), UL:(1.99, 2.39), LR:(2, 1.6)

Fig. 9.91 Equation (9.48): $b = 1.6$, $k = m = n = \gamma = \mu = 1$, $\alpha = 1.5$, $\beta = 0$, B versus A, r:AABB AABB..., $n_{\text{prev}} = 100$, $n_{\max} = 200$, $x_0 = 0.3$. D-shading. LL:(2, 3.15), UL:(3.15, 4.3), LR:(3.15, 2)

Fig. 9.92 Equation (9.48): $b = 2.5$, $k = m = n = \gamma = \mu = 1$, $\alpha = 0$, $\beta = 0.5$, B versus A, r:AB AB..., $n_{\text{prev}} = 400$, $n_{\max} = 1000$, $x_0 = 0.5$. D-shading. LL:(1.695, 3.125), UL:(2.455, 3.885), LR:(3.125, 1.695)

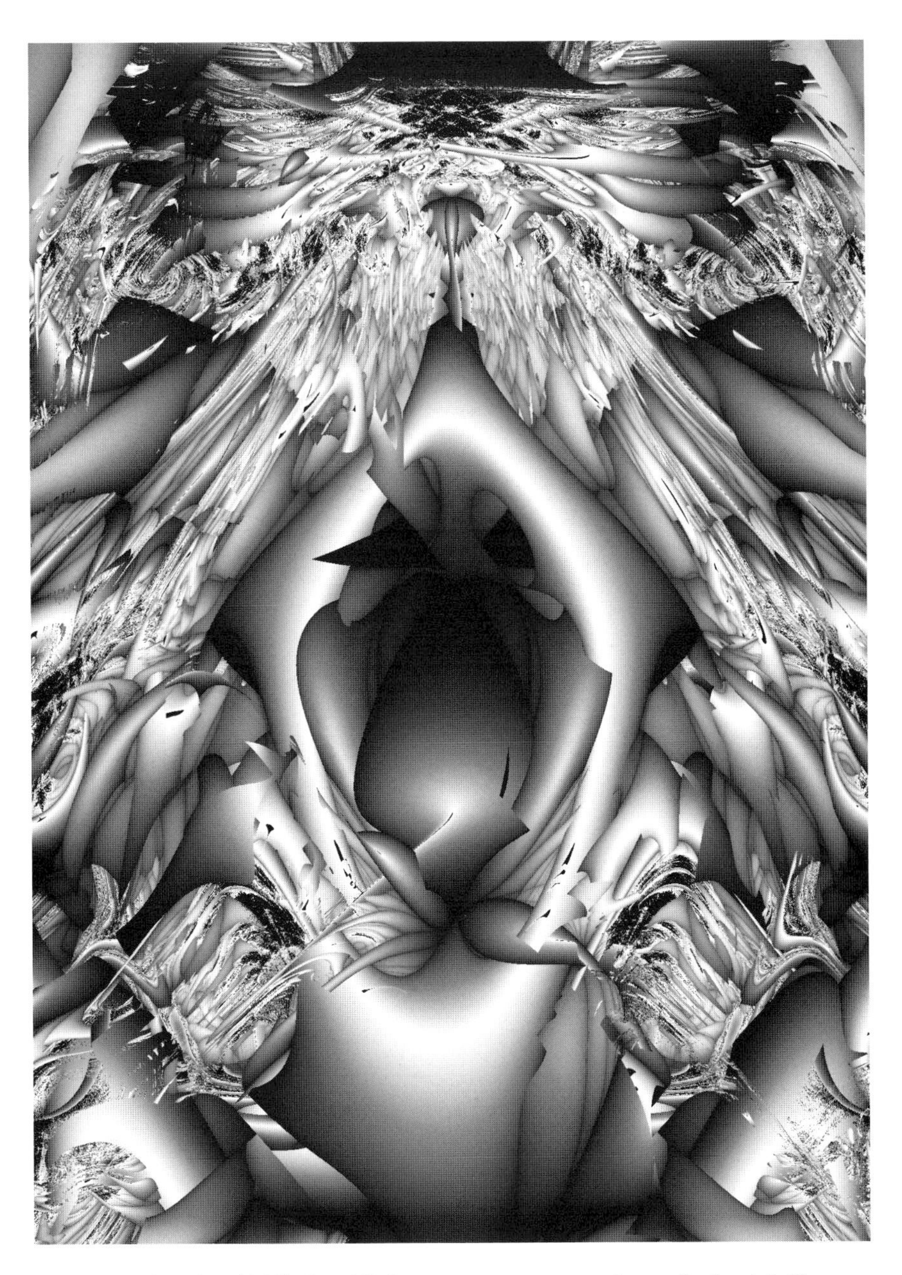

Fig. 9.93 Equation (9.48): $b = 1.7$, $k = m = n = \gamma = \mu = 1$, $\alpha = 0$, $\beta = 1.5$, B versus A, r:AAABBB AAABBB..., $n_{\text{prev}} = 100$, $n_{\text{max}} = 200$, $x_0 = 0.3$. D-shading. LL:(1.59, 2.37), UL:(2.545, 3.325), LR:(2.36, 1.6)

Fig. 9.94 Equation (9.48): $b = 2.36$, $k = m = n = \gamma = \mu = 1$, $\alpha = 0$, $\beta = 0.16$, B versus A, r:A^7B^7 A^7B^7..., $n_{\text{prev}} = 100$, $n_{\text{max}} = 200$, $x_0 = 0.5$. D-shading. LL:(10.155, 10.6), UL:(10.5085, 10.9535), LR:(10.60, 10.155)

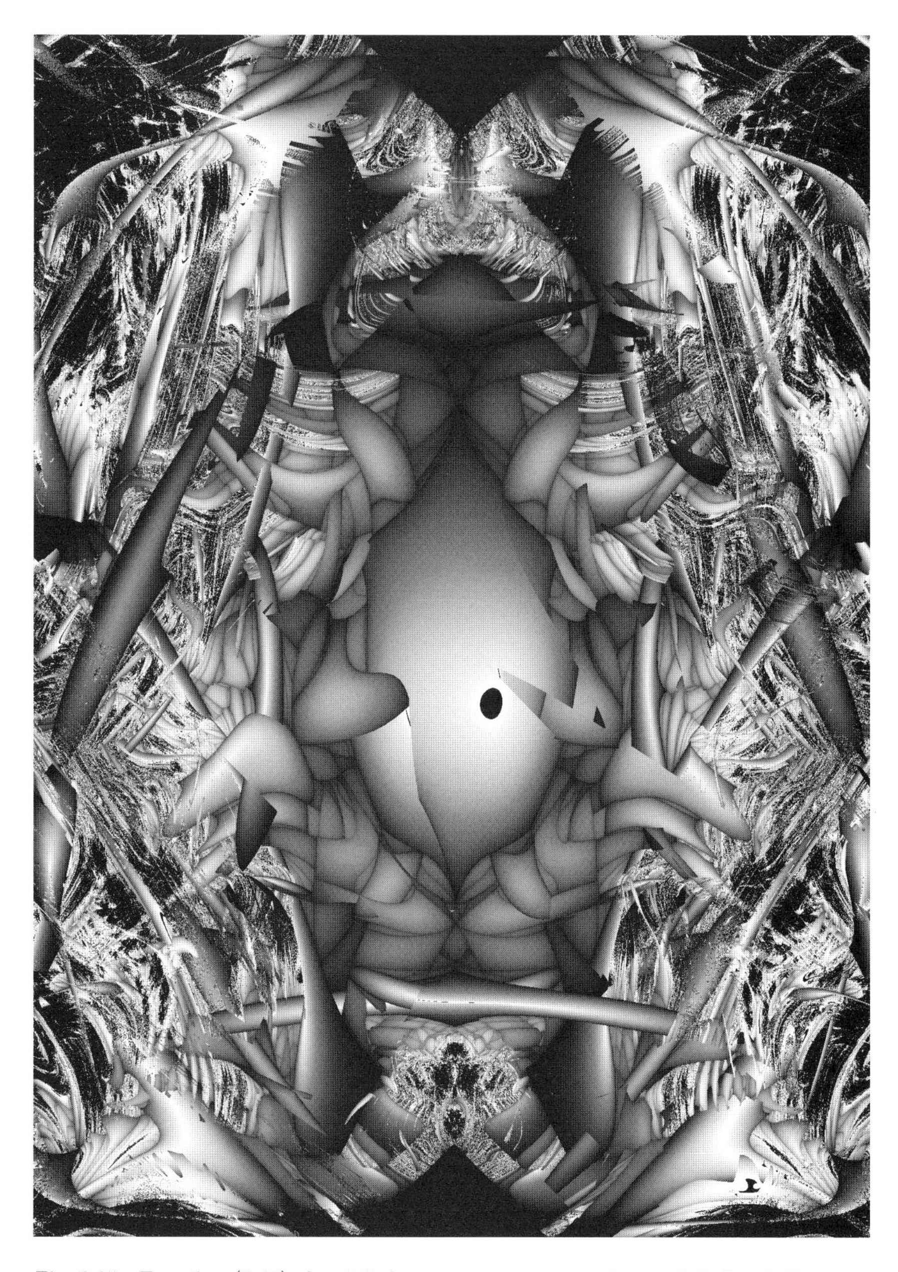

Fig. 9.95 Equation (9.48): $b = 2.3$, $k = m = n = \gamma = \mu = 1$, $\alpha = 0.2$, $\beta = 0$, B versus A, r:AAABBB AAABBB..., $n_{\text{prev}} = 100$, $n_{\max} = 200$, $x_0 = 0.3$. D-shading. LL:(6.928, 7.448), UL:(7.65, 8.18), LR:(7.448, 6.935)

Fig. 9.96 Equation (9.48): $b = 2$, $k = m = n = 1$, $\gamma = \mu = 2$, $\alpha = 0.7$, $\beta = 0$, B versus A, r:AABB AABB..., $n_{\text{prev}} = 800$, $n_{\max} = 500$, $x_0 = 0.5$. D-shading. LL:(−0.912, −1.012), UL:(−0.994, −1.094), LR:(−1.012, −0.912)

Fig. 9.97 Equation (9.48): $b = 2$, $k = m = n = 1$, $\gamma = \mu = 2$, $\alpha = 0$, $\beta = 1.25$, B versus A, r:AAABBB AAABBB..., $n_{\text{prev}} = 1000$, $n_{\text{max}} = 1000$, $x_0 = 1.5$. D-shading. LL:(5.1283, 4.9518), UL:(5.2612, 5.1886), LR:(5.345, 4.8302)

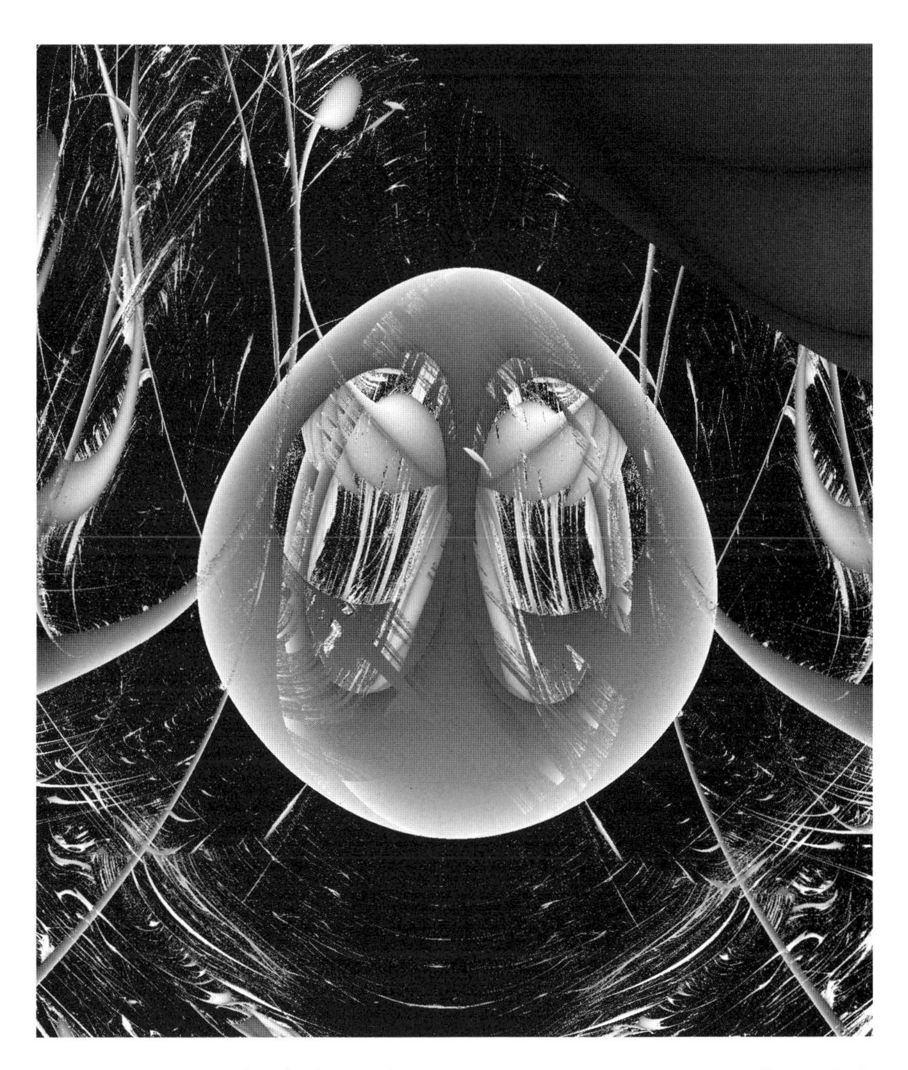

Fig. 9.98 Equation (9.48): $b = 2$, $k = m = n = 1$, $\gamma = \mu = 2$, $\alpha = 0$, $\beta = 1.25$, B versus A, r:AAABBB AAABBB..., $n_{\text{prev}} = 3000$, $n_{\max} = 3000$, $x_0 = 1.5$. D-shading. LL:(-0.778, -0.5766), UL:(-0.557, -0.3296), LR:(-0.5544, -0.77666)

Fig. 9.99 Equation (9.48): $b = 2$, $k = m = n = 1$, $\gamma = \mu = 2$, $\alpha = 0$, $\beta = 1.25$, B versus A, r:AAABBB AAABBB..., $n_{\text{prev}} = 500$, $n_{\text{max}} = 500$, $x_0 = 1.5$. D-shading. LL:(5.84, 5.44), UL:(5.35, 4.95), LR:(5.44, 5.84)

Fig. 9.100 Equation (9.48): $b = 2$, $k = m = n = 1$, $\gamma = \mu = 2$, $\alpha = 0$, $\beta = 1.25$, B versus A, r:AAABBB AAABBB..., $n_{\text{prev}} = 500$, $n_{\max} = 500$, $x_0 = 1.5$. D-shading. LL:(3.56, 3.881), UL:(3.56, 4.75), LR:(3.829, 3.881)

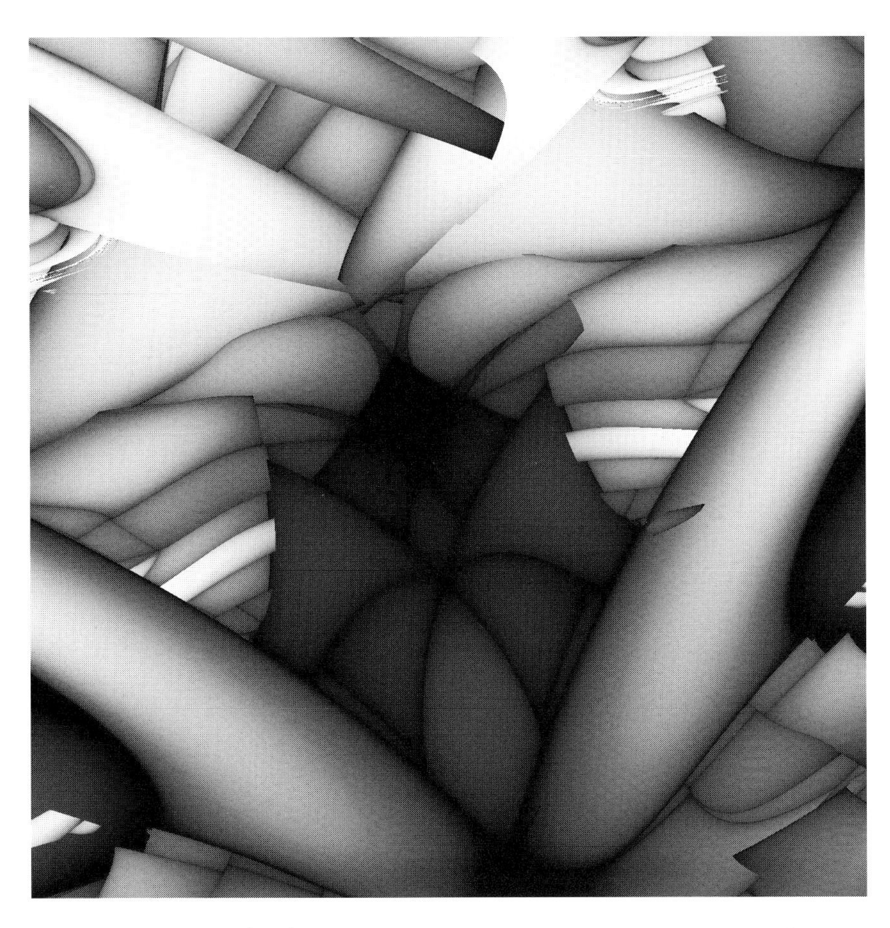

Fig. 9.101 Equation (9.48): $b = 1.5$, $\gamma = \mu = k = 1$, $m = n = 2$, $\alpha = 0.6$, $\beta = 0$, B versus A, r:AABB AABB..., $n_{\text{prev}} = 100$, $n_{\max} = 200$, $x_0 = 0.5$. D-shading. LL:(3.2199, 3.3405), UL:(3.338, 3.4105), LR:(3.297, 3.2104)

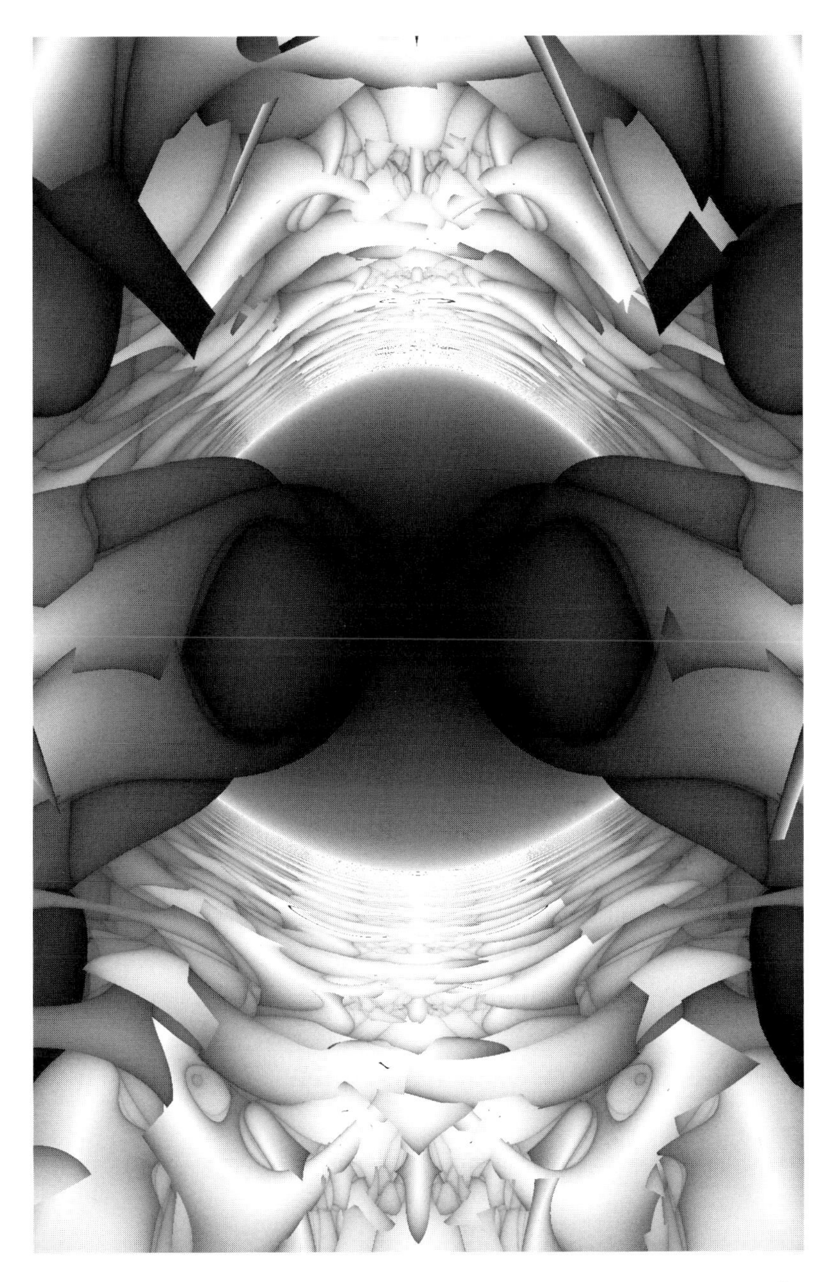

Fig. 9.102 Equation (9.48): $b = 1.5$, $\gamma = \mu = k = 1$, $m = n = 2$, $\alpha = -1.25$, $\beta = 0$, B versus A, r:AABB AABB..., $n_{\text{prev}} = 1000$, $n_{\max} = 1000$, $x_0 = 0.6$. D-shading. LL:(0.784, 1.216), UL:(1.3, 1.732), LR:(1.216, 0.784)

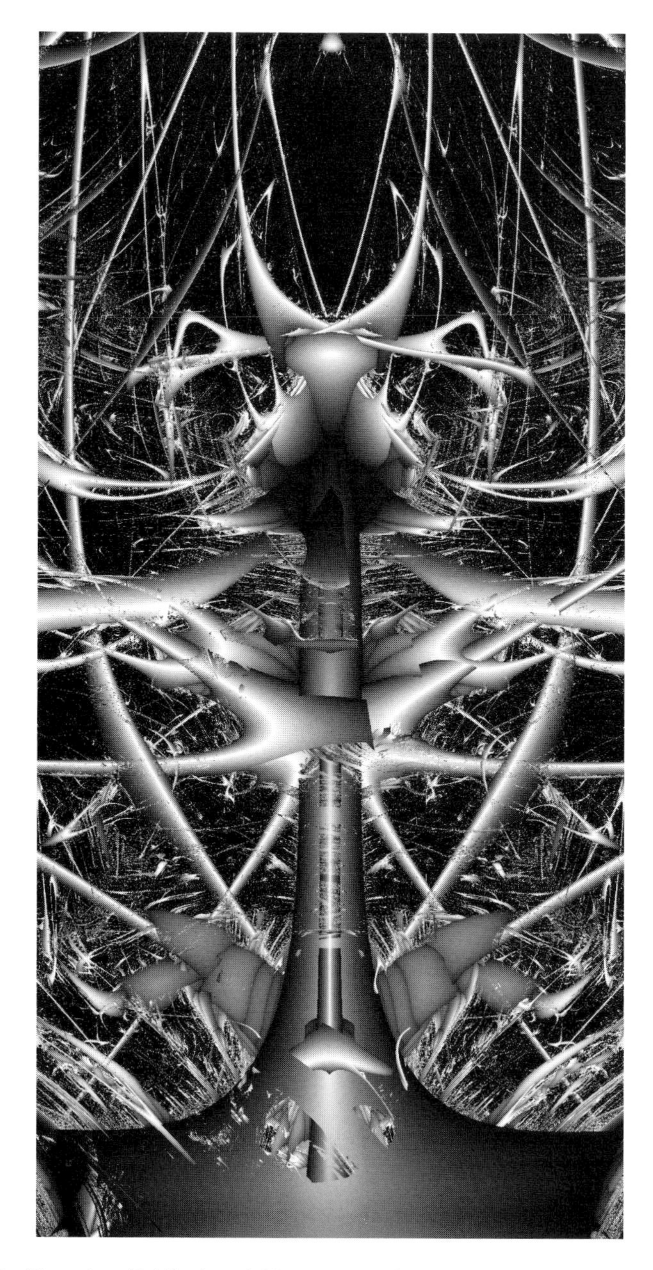

Fig. 9.103 Equation (9.48): $b = 2.25$, $\gamma = \mu = k = 1$, $m = n = 2$, $\alpha = -1$, $\beta = 0$, B versus A, r:AABB AABB..., $n_{\text{prev}} = 2000$, $n_{\text{max}} = 2000$, $x_0 = 0.5$. D-shading. LL:(5.6905, 5.73), UL:(5.7505, 5.79), LR:(5.73, 5.6905)

Fig. 9.104 Equation (9.48): $b = 2.25$, $\gamma = \mu = k = 1$, $m = n = 2$, $\alpha = -1$, $\beta = 0$, B versus A, r:AABB AABB..., $n_{\text{prev}} = 100$, $n_{\max} = 200$, $x_0 = 0.5$. D-shading. LL:(1.171, 1.499), UL:(1.659, 1.977), LR:(1.489, 1.1743)

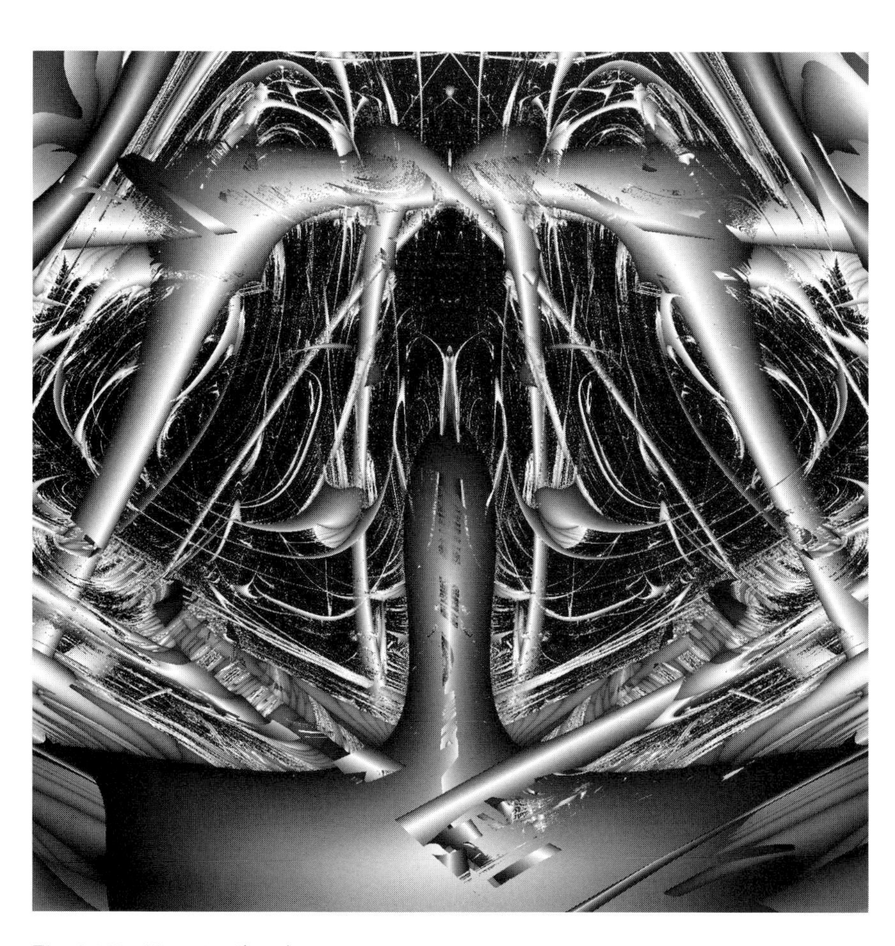

Fig. 9.105 Equation (9.48): $b = 2$, $\gamma = \mu = k = 1$, $m = n = 2$, $\alpha = 0.7$, $\beta = 0$, B versus A, r:AABB AABB..., $n_{\text{prev}} = 2000$, $n_{\max} = 2000$, $x_0 = 0.5$. D-shading. LL:(4.8036, 4.8856), UL:(4.8952, 4.9772), LR:(4.8856, 4.8036)

Fig. 9.106 Equation (9.48): $b = 1.6$, $\gamma = \mu = k = 1$, $m = n = 2$, $\alpha = 0$, $\beta = 1.5$, B versus A, r:AABB AABB..., $n_{\text{prev}} = 100$, $n_{\max} = 200$, $x_0 = 0.3$. D-shading. LL:(5.268, 5.456), UL:(5.198, 5.901), LR:(5.43, 5.48148)

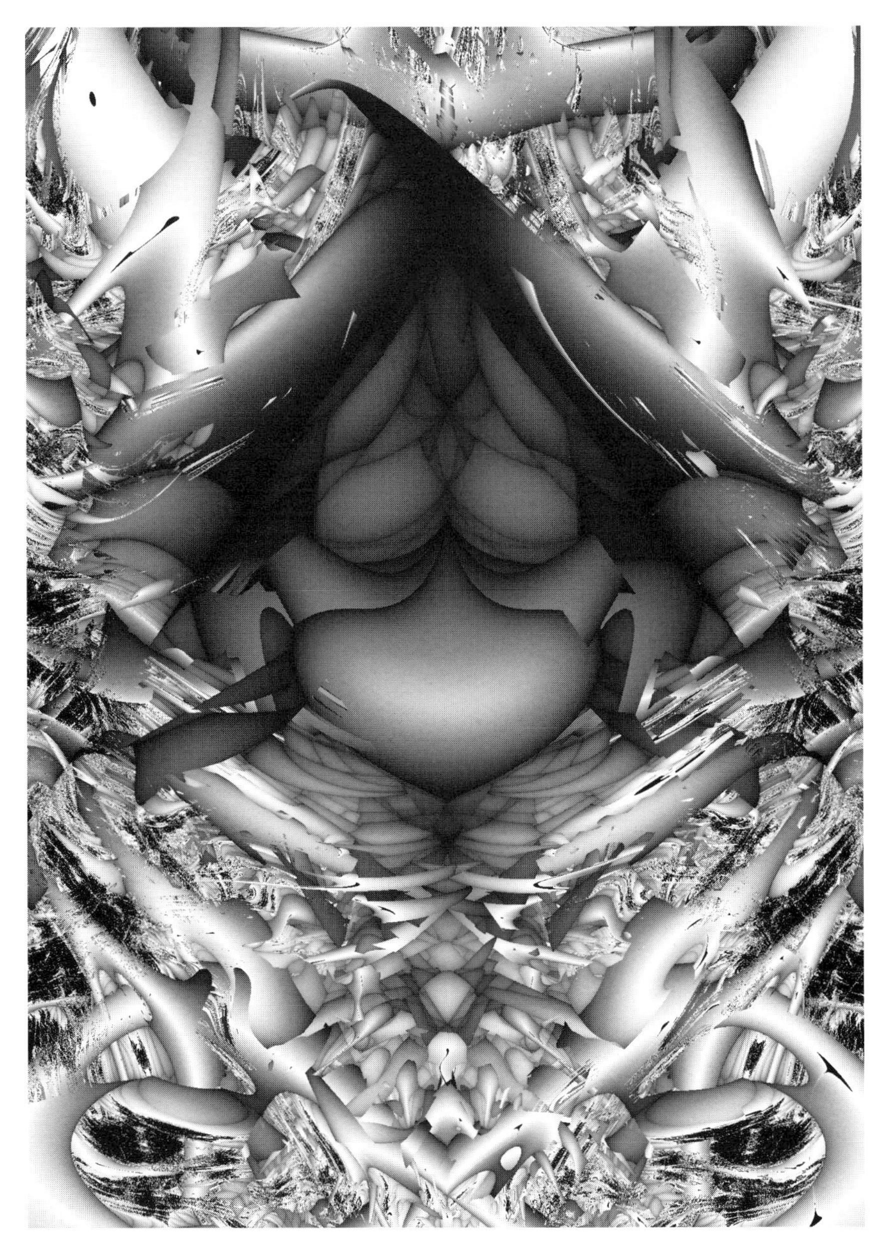

Fig. 9.107 Equation (9.48): $b = 1.8$, $\gamma = \mu = k = 1$, $m = n = 2$, $\alpha = 0$, $\beta = 1.5$, B versus A, r:AAABBB AAABBB..., $n_{\text{prev}} = 100$, $n_{\text{max}} = 200$, $x_0 = 0.5$. D-shading. LL:(3.363, 3.499), UL:(3.498, 3.635), LR:(3.499, 3.364)

Fig. 9.108 Equation (9.48): $b = 1.8$, $\gamma = \mu = k = 1$, $m = n = 2$, $\alpha = 0$, $\beta = 1.5$, B versus A, r:AAABBB AAABBB..., $n_{\text{prev}} = 100$, $n_{\max} = 200$, $x_0 = 0.5$. D-shading. LL:(2.257, 2.467), UL:(2.495, 2.705), LR:(2.467, 2.257)

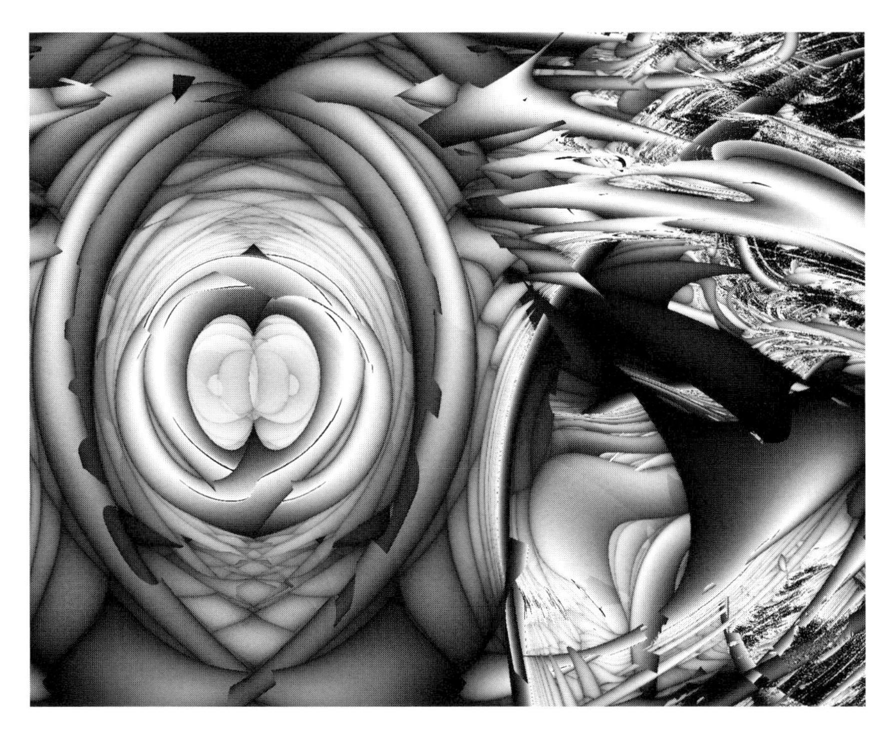

Fig. 9.109 Equation (9.48): $b = 1.8$, $\gamma = \mu = k = 1$, $m = n = 2$, $\alpha = 0$, $\beta = 1.5$, B versus A, r:AAABBB AAABBB..., $n_{\text{prev}} = 200$, $n_{\text{max}} = 200$, $x_0 = 0.5$. D-shading. LL:(0.304, 0.458), UL:(0.552, 0.706), LR:(0.59, 0.172)

Fig. 9.110 Equation (9.48): $b = 2$, $\gamma = \mu = k = 1$, $m = n = 2$, $\alpha = -0.4$, $\beta = 0$, B versus A, r:AABB AABB..., $n_{\text{prev}} = 200$, $n_{\max} = 200$, $x_0 = 0.3$. D-shading. LL:(5.076, 5.206), UL:(5.181, 5.311), LR:(5.206, 5.076)

Fig. 9.111 Equation (9.48): $b = 2$, $\gamma = \mu = k = 1$, $m = n = 2$, $\alpha = -0.4$, $\beta = 0$, B versus A, r:AABB AABB..., $n_{\text{prev}} = 100$, $n_{\max} = 200$, $x_0 = 0.3$. D-shading. LL:(2.164, 2.335), UL:(2.425, 2.596), LR:(2.335, 2.164)

Fig. 9.112 Equation (9.48): $b = 2$, $\gamma = \mu = k = 1$, $m = n = 2$, $\alpha = -0.4$, $\beta = 0$, B versus A, r:AABB AABB..., $n_{\text{prev}} = 200$, $n_{\max} = 200$, $x_0 = 0.3$. D-shading. LL:(0.829, 1.099), UL:(1.342, 1.612), LR:(1.099, 0.829)

Fig. 9.113 Equation (9.48): $b = 1.7$, $\gamma = \mu = k = 1$, $m = n = 2$, $\alpha = 0$, $\beta = 1.5$, B versus A, r:AABB AABB..., $n_{\text{prev}} = 100$, $n_{\max} = 200$, $x_0 = 0.3$. D-shading. LL:(4.7475, 4.774), UL:(4.7888, 4.8198), LR:(4.7785, 4.7460)

Fig. 9.114 Equation (9.48): $b = 2$, $\mu = k = m = n = 1$, $\gamma = 2$, $\alpha = 0$, $\beta = 1.25$, B versus A, r:AAABBB AAABBB..., $n_{\text{prev}} = 100$, $n_{\max} = 200$, $x_0 = 1.5$. D-shading. LL:(1.988, 2.796), UL:(3.258, 4.066), LR:(2.796, 1.988)

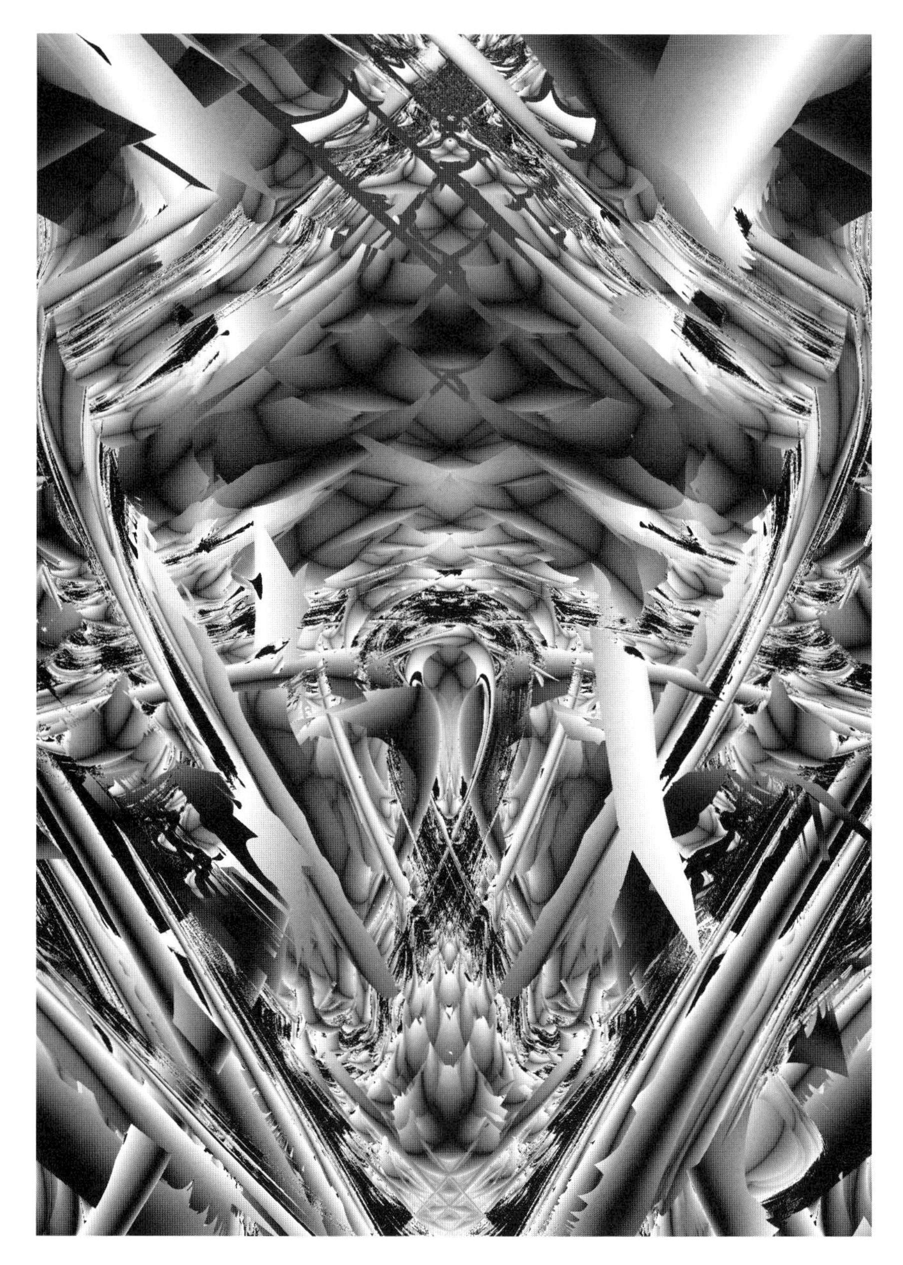

Fig. 9.115 Equation (9.48): $b = 1.6$, $\gamma = \mu = k = m = 1$, $n \to -\infty$, $\alpha = 0$, $\beta = -2.8$, B versus A, r:AABB AABB..., $n_{\text{prev}} = 100$, $n_{\max} = 200$, $x_0 = 0.39$. D-shading. LL:(6.78, 7.12), UL:(7.41, 7.75), LR:(7.123, 6.777)

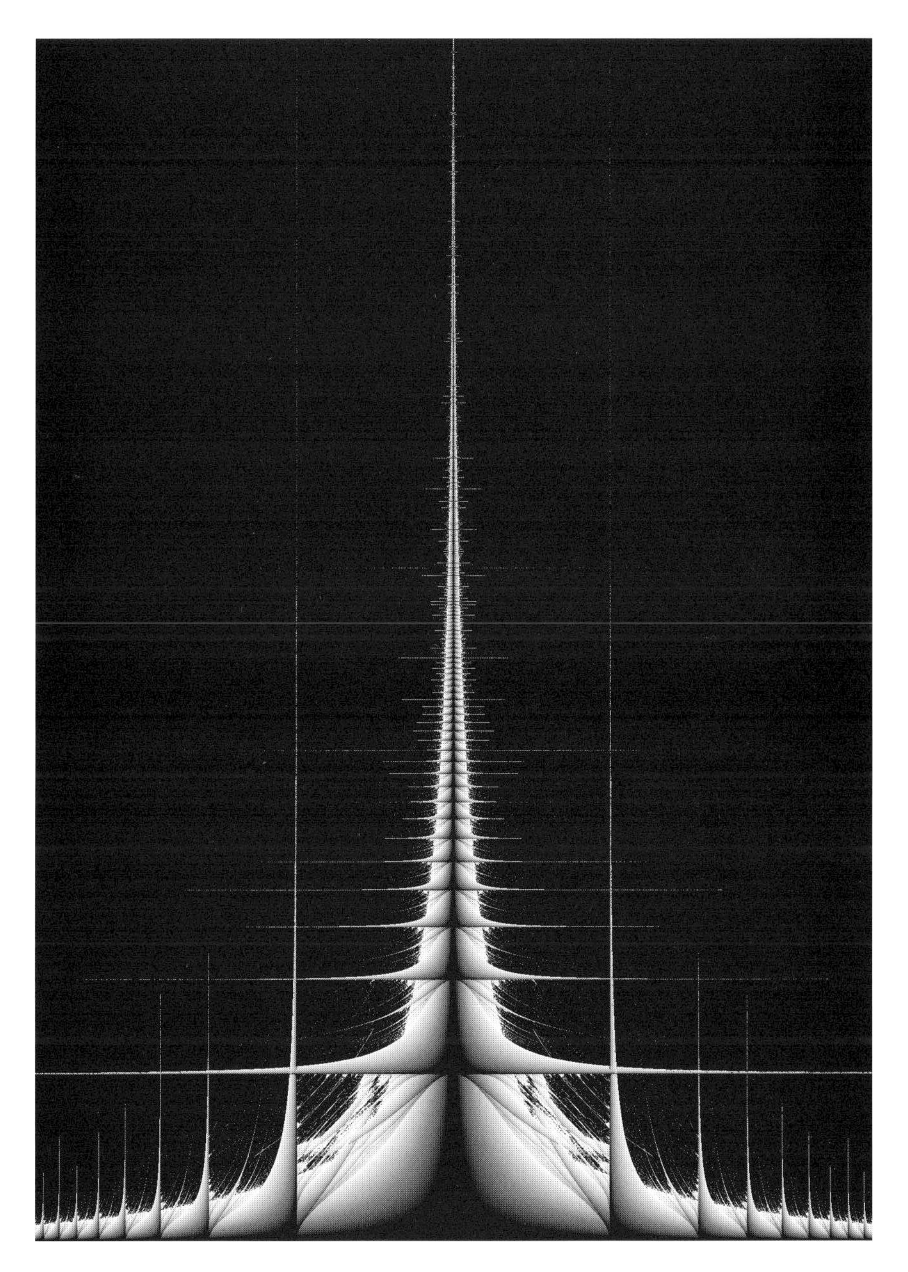

Fig. 9.116 Equation (9.49): $b = 1.9$, B versus A, r:AB AB..., $n_{\rm prev} = 100$, $n_{\rm max} = 200$, $x_0 = 1.5$. L-shading. LL:(-1.86, 0), UL:(-1.86, 5), LR:(1.86, 0)

Fig. 9.117 Equation (9.50): $b = 1.9$, B versus A, r:AB AB..., $n_{\text{prev}} = 100$, $n_{\text{max}} = 200$, $x_0 = 1.5$. D-shading. LL:$(-3.8, 0)$, UL:$(-3.8, 6)$, LR:$(3.8, 0)$

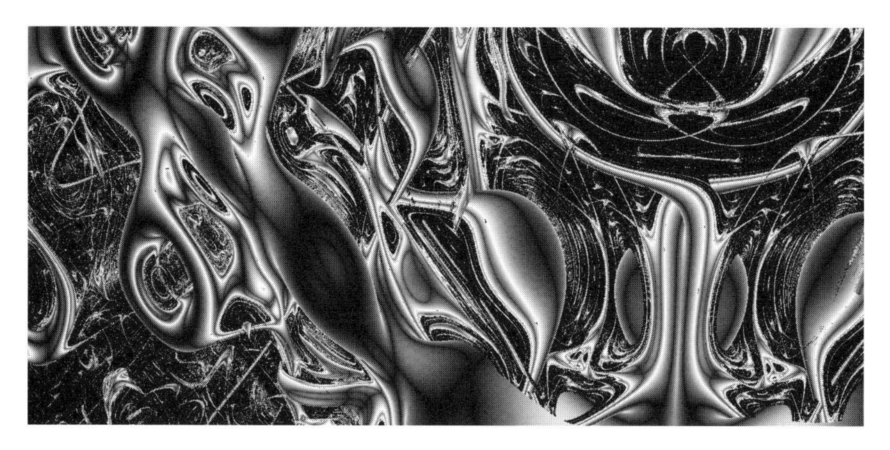

Fig. 9.118 Equation (9.51): $b = 0.6$, B versus A, r:AB AB..., $n_{\text{prev}} = 100$, $n_{\text{max}} = 100$, $x_0 = 0.5$. D-shading. LL:$(-0.9538, 1.8213)$, UL:$(-0.1675, 2.7)$, LR:$(0.9425, 0.1246)$

Fig. 9.119 Equation (9.52): $b = 3.2$, B versus A, r:AB AB..., $n_{\text{prev}} = 25$, $n_{\max} = 50$, $x_0 = 0.5$. L-shading. LL:(−1.52, 1.33), UL:(1.33, 4.18), LR:(1.33, −1.52)

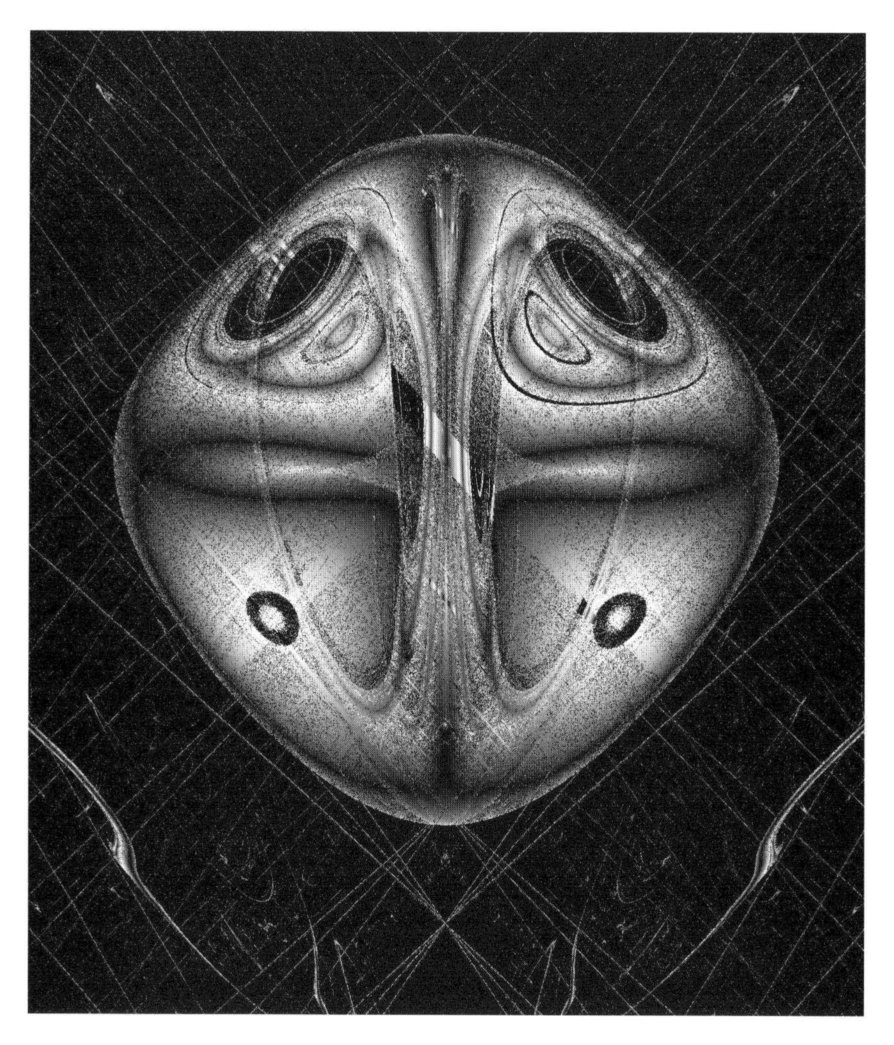

Fig. 9.120 Equation (9.53): $b = 4$, B versus A, r:AAABBB AAABBB..., $n_{\text{prev}} = 400$, $n_{\max} = 400$, $x_0 = 0.5$. D-shading. LL:(1.0913, 1.2253), UL:(1.2444, 1.3784), LR:(1.2253, 1.0913)

Fig. 9.121 Equation (9.54): $b = 1.5$, B versus A, r:AABABAB AABABAB..., $n_{\text{prev}} = 500$, $n_{\max} = 1000$, $x_0 = 0.5$. D-shading. LL:(−6, −5), UL:(−6, 10.36), LR:(6, −5)

Fig. 9.122 Equation (9.55): $b = 0.8$, B versus A, r:AB AB..., $n_{\text{prev}} = 100$, $n_{\text{max}} = 500$, $x_0 = 0.5$. L-shading. LL:(−5, 3), UL:(3, 11), LR:(3, −5)

Fig. 9.123 Equation (9.56): $b = 1$, B versus A, r:AABB AABB..., $n_{\text{prev}} = 100$, $n_{\text{max}} = 200$, $x_0 = 0.5$. L-shading. LL:(−10, −1.6666), UL:(−10, 0.04), LR:(−6.6666, −1.6666)

Fig. 9.124 Equation (9.57): $b = 1.2$, B versus A, r:AAABB AAABB..., $n_{\text{prev}} = 300$, $n_{\text{max}} = 600$, $x_0 = 0.5$. L-shading. LL:(0.5, 0.2), UL:(0.5, 20), LR:(20, 0.2)

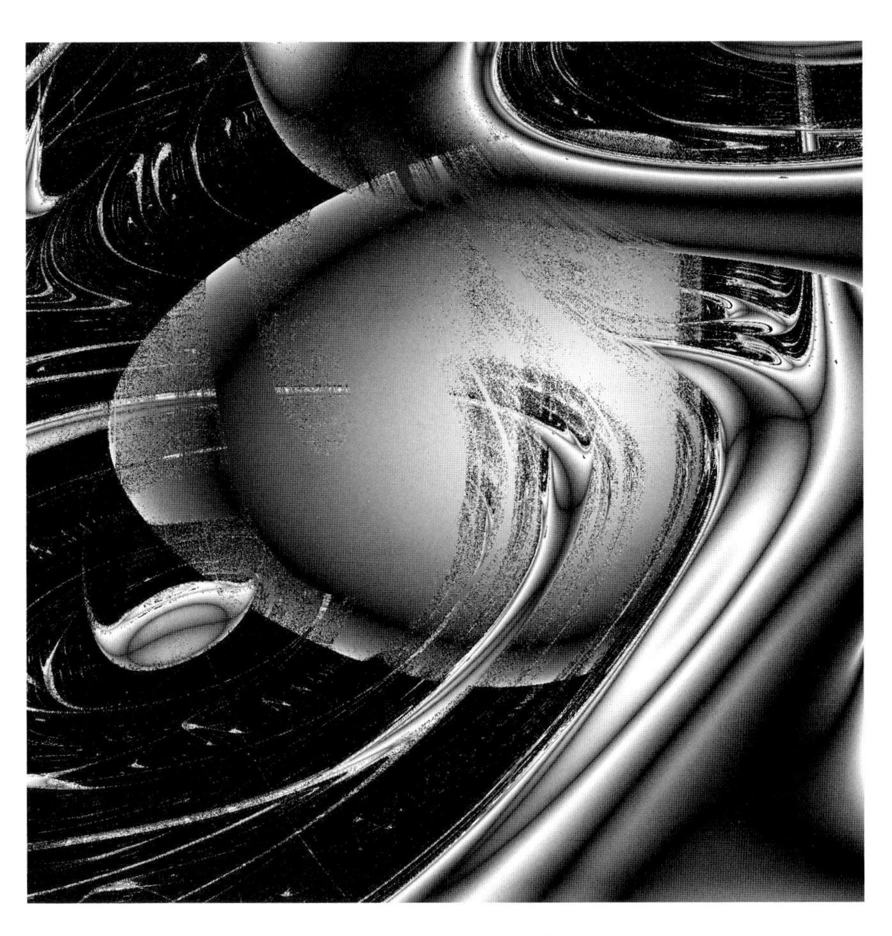

Fig. 9.125 Equation (9.58): $b = 1.1$, B versus A, r:AB AB..., $n_{\text{prev}} = 50$, $n_{\max} = 100$, $x_0 = 0.9$. D-shading. LL:(−0.64, 1.76), UL:(−0.64, 2.1), LR:(−0.3, 1.76)

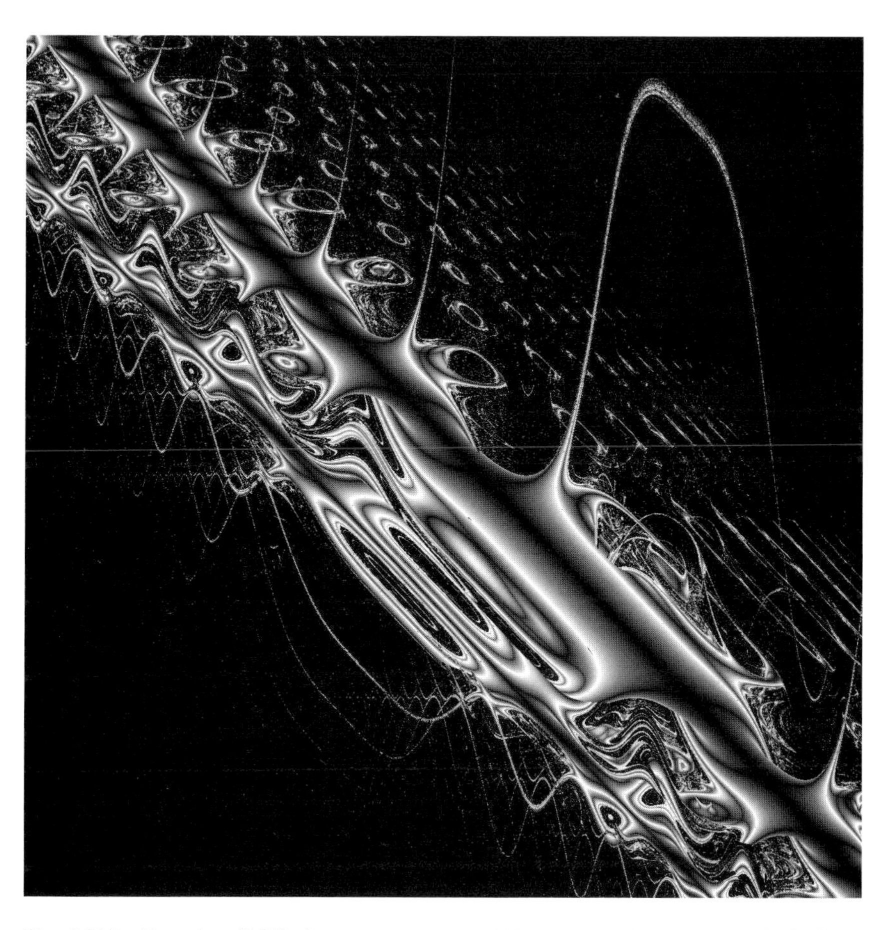

Fig. 9.126 Equation (9.59): b versus r, $n_{\text{prev}} = 100$, $n_{\max} = 200$, $x_0 = 0.5$. D-shading. LL:(−5.7, −2.5), UL:(−5.7, 7.2), LR:(4.3, −2.5)

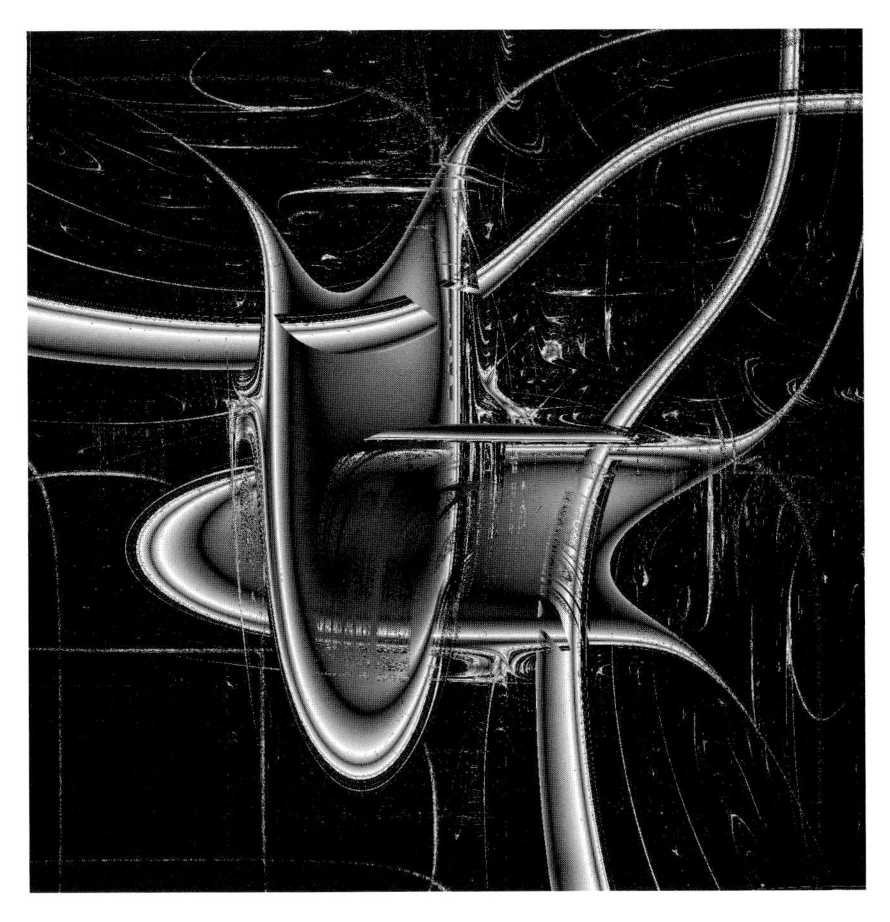

Fig. 9.127 Equation (9.60): $b = 1$, B versus A, r:AB AB..., $n_{\text{prev}} = 150$, $n_{\max} = 150$, $x_0 = -0.5$. D-shading. LL:(0.2, 0.2), UL:(0.2, 2), LR:(2, 0.2)

Fig. 9.128 Equation (9.58): $b = 0.8$, B versus A, r:AB AB..., $n_{\text{prev}} = 25$, $n_{\text{max}} = 50$, $x_0 = 0.5$. D-shading. LL:(1.02, 2.2), UL:(1.02, 3.1), LR:(2.06, 2.2)

Fig. 9.129 Equation (9.61): $b = 5$, B versus A, r:AAAAABBBBB AAAAABBBBB..., $n_{\text{prev}} = 25$, $n_{\text{max}} = 50$, $x_0 = 0.5$. L-shading. LL:(1.3341, 4.7792), UL:(3.2843, 6.7294), LR:(4.7795, 1.3338)

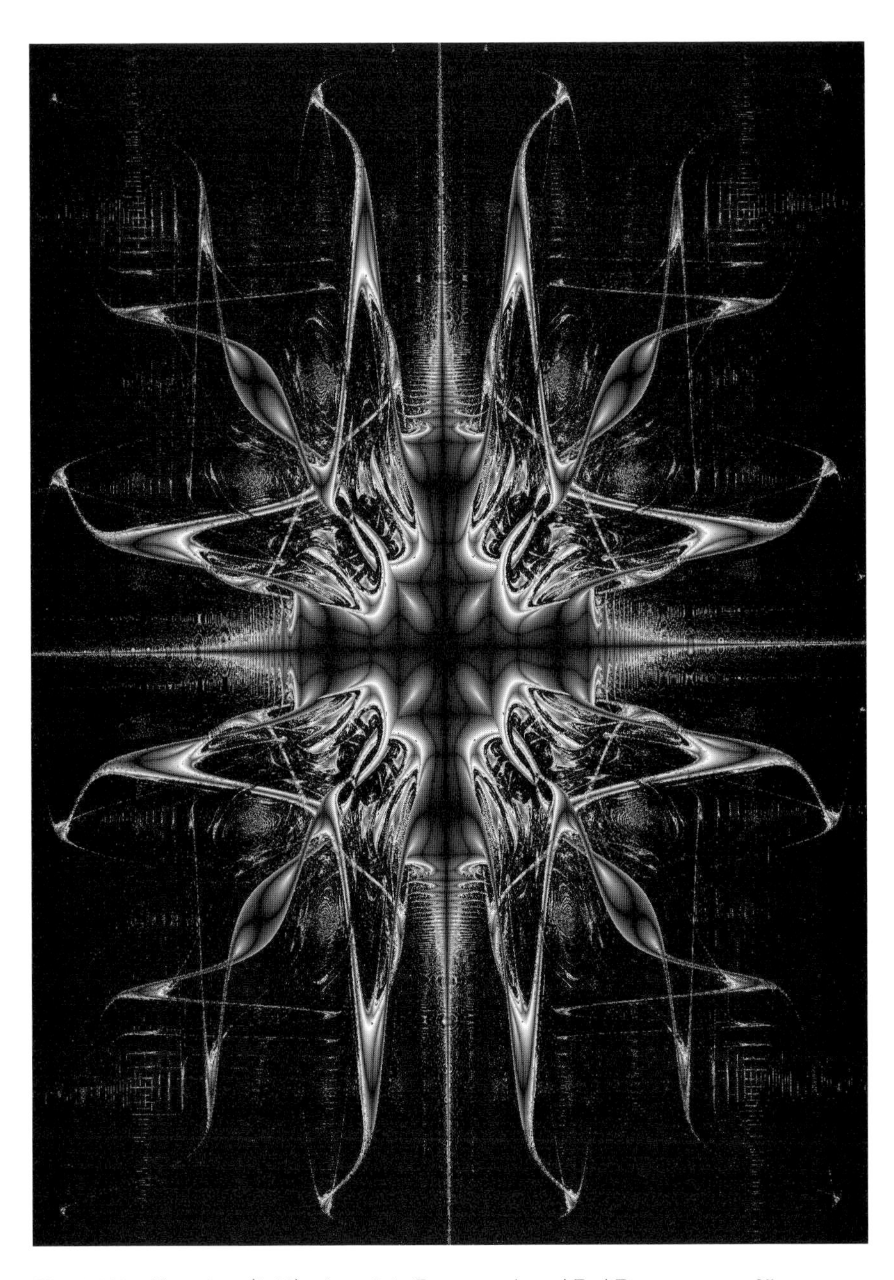

Fig. 9.130 Equation (9.62): $b = 0.6$, B versus A, r:AB AB..., $n_{\text{prev}} = 25$, $n_{\max} = 50$, $x_0 = 0.5$. D-shading. LL:(−4.0736, −4.0119), UL:(−4.0119, 4.1403), LR:(4.0786, −4.0736)

Fig. 9.131 Equation (9.63): $b = 0.75$, B versus A, r:AB AB..., $n_{\text{prev}} = 300$, $n_{\max} = 500$, $x_0 = 0.5$. D-shading. LL:(7.4242, −10), UL:(7.4242, 10), LR:(−8.6742, −10)

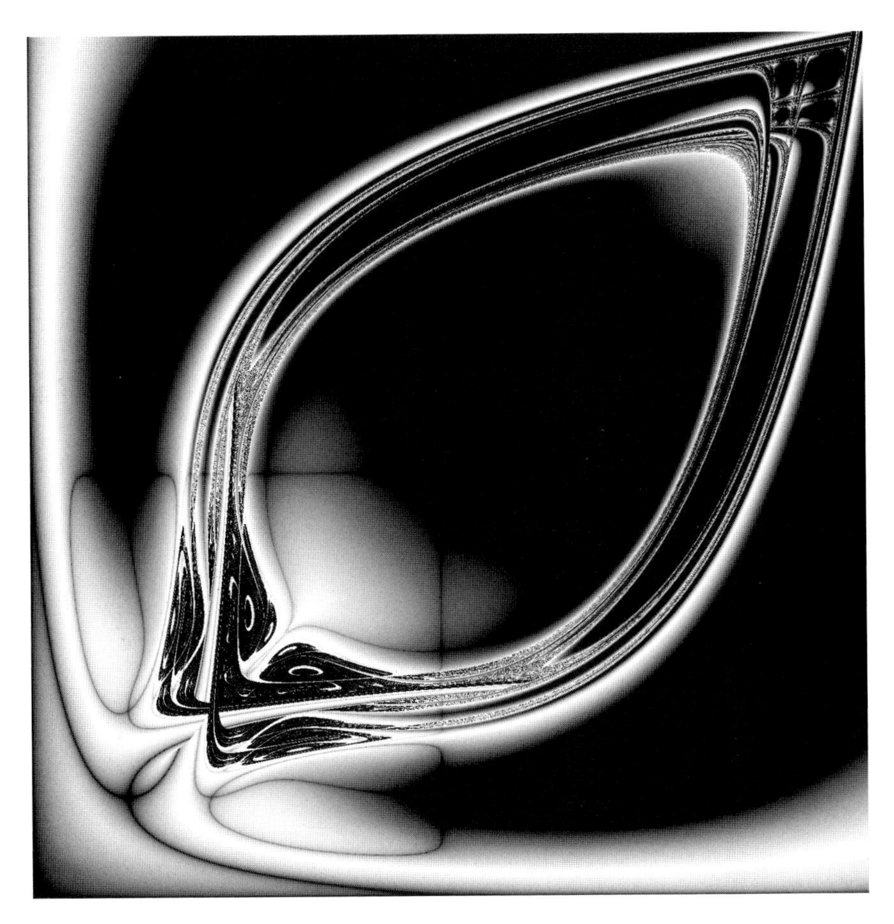

Fig. 9.132 Equation (9.64): $b = 1.2$, B versus A, r:AB AB..., $n_{\text{prev}} = 500$, $n_{\text{max}} = 2000$, $x_0 = 0.7$. L-shading($\lambda_m = -2$). LL:(0, 0), UL:(0, 10.3), LR:(10.3, 0)

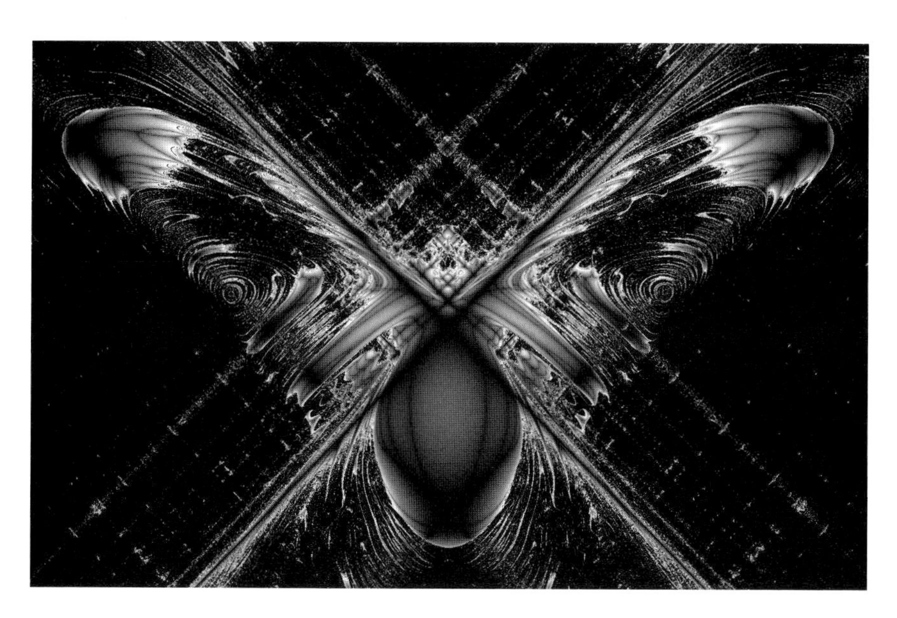

Fig. 9.133 Equation (9.65): $b = 1$, B versus A, r:AAAABBBB AAAABBBB..., $n_{\text{prev}} = 500$, $n_{\text{max}} = 500$, $x_0 = 0.5$. D-shading. LL:(−15.555, 3.8982), UL:(−3.1422, 16.126), LR:(3.7129, −15.6612)

Fig. 9.134 Equation (9.66): $b = 1$, B versus A, r:AAABBB AAABBB..., $n_{\text{prev}} = 100$, $n_{\text{max}} = 200$, $x_0 = 0.5$. L-shading. LL:(−2, 3), UL:(1.642, 6.642), LR:(3, −2)

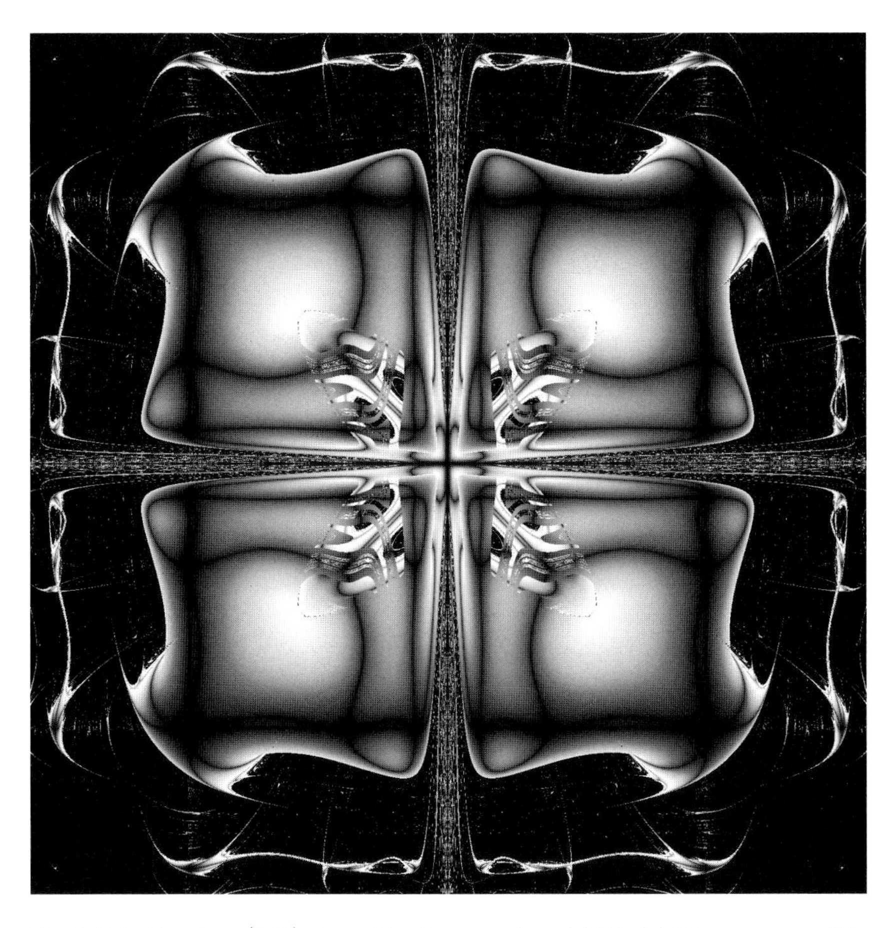

Fig. 9.135 Equation (9.67): $b = 0.1$, B versus A, r:AABB AABB..., $n_{\text{prev}} = 200$, $n_{\max} = 200$, $x_0 = 0.5$. D-shading. LL:$(-2.8, -2.8)$, UL:$(-2.8, 2.8)$, LR:$(2.8, -2.8)$

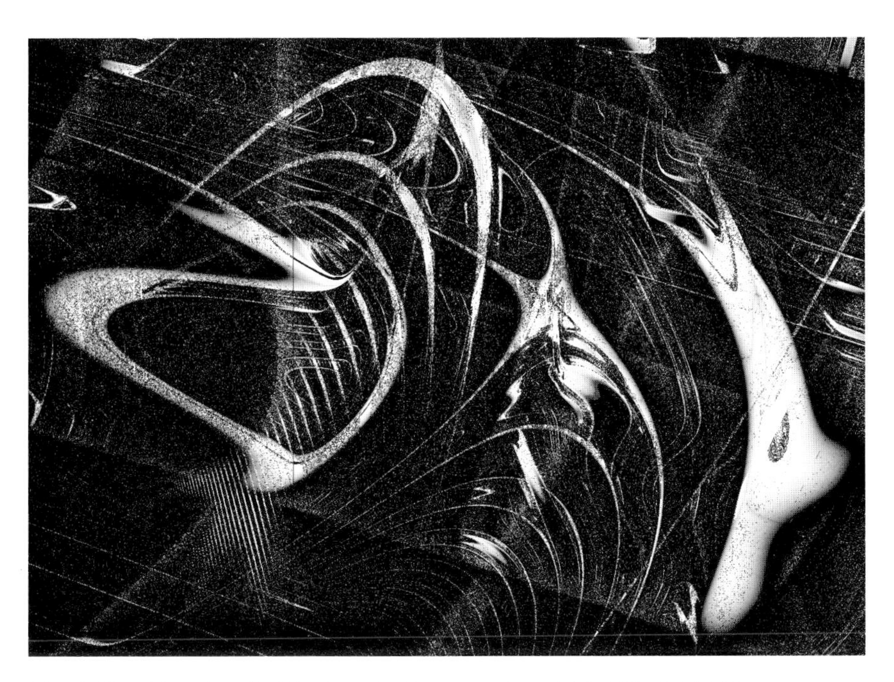

Fig. 9.136 Equation (9.68): $b = 1$, B versus A, r:AAABBB AAABBB..., $n_{\text{prev}} = 100$, $n_{\text{max}} = 200$, $x_0 = 0.9$. L-shading. LL:(2.21248, 1.53388), UL:(2.21248, 1.61899), LR:(2.33097, 1.53388)

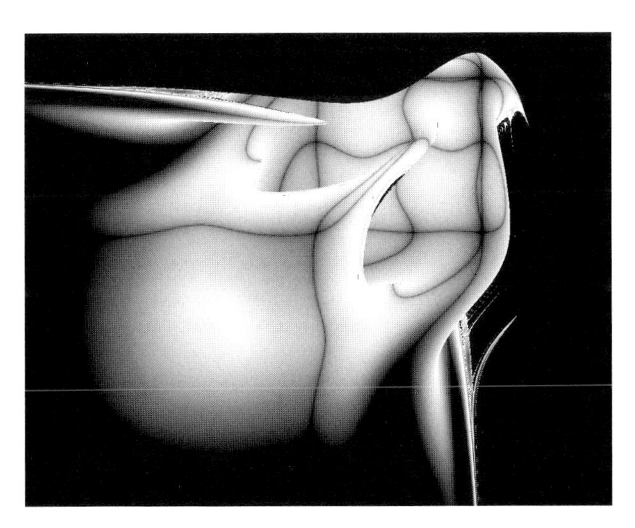

Fig. 9.137 Equation (9.69): $b = 2$, B versus A, r:ABBA ABBA..., $n_{\text{prev}} = 200$, $n_{\text{max}} = 100$, $x_0 = 1.135$. L-shading. LL:(−2.3484, −2.3484), UL:(−2.3484, 0.5785), LR:(1.05, −2.3484)

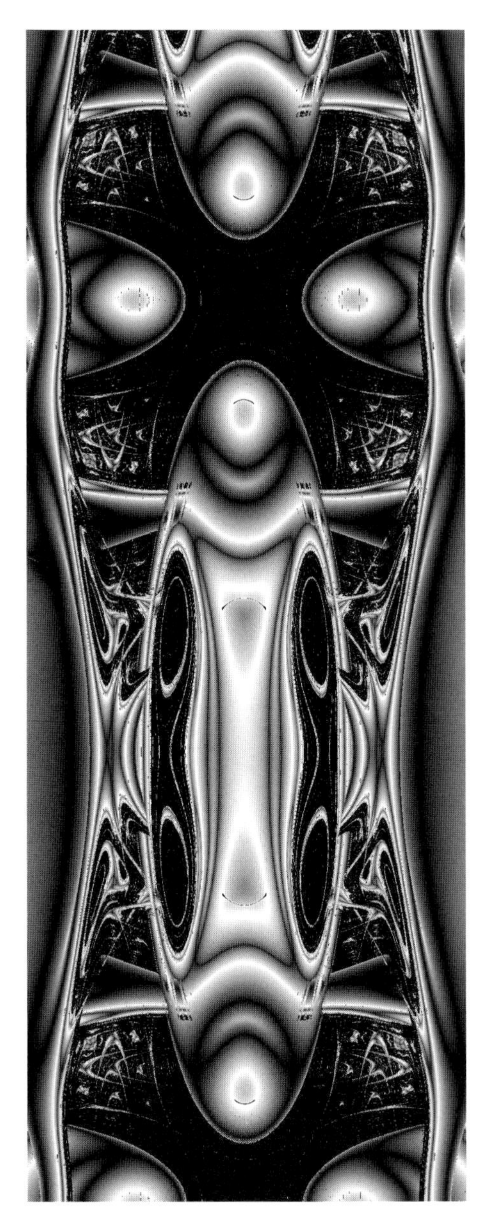

Fig. 9.138 Equation (9.70): b = 1.5, B versus A, r:AB AB..., n_{prev} = 50, n_{max} = 100, x_0 = 0.5. D-shading. LL:(6.2879, −4.697), UL:(6.2879, 3.447), LR:(9.4334, −4.697)

Fig. 9.139 Equation (9.71): b = 1.5, B versus A, r:AAABBB AAABBB..., n_{prev} = 25, n_{max} = 50, x_0 = 0.5. L-shading. LL:(−2.7, −1.08), UL:(1.08, 2.7), LR:(−1.08, −2.7)

Fig. 9.140 Equation (9.72): $b = 0.8$, B versus A, r:AB AB..., $n_{\rm prev} = 25$, $n_{\rm max} = 50$, $x_0 = 0.5$. D-shading. LL:(−5.4256, 1.7232), UL:(−3.397, 10.2192), LR:(1.5161, −5.4853)

Fig. 9.141 Equation (9.73): $b = 1.1$, B versus A, r:AB AB..., $n_{\text{prev}} = 125$, $n_{\max} = 250$, $x_0 = 0.5$. D-shading. LL:(−2.0185, −1.6632), UL:(−1.7052, −1.0454), LR:(−1.3325, −2.0111)

Fig. 9.142 Equation (9.74): b = 1.8, B versus A, r:A^{6}BB A^{6}BB..., n_{prev} = 100, $n_{\max}$ = 200, x_0 = 0.5. D-shading. LL:(−2.62, 0.36), UL:(−2.62, 26.04), LR:(2.6, 0.36)

Fig. 9.143 Equation (9.75): $b = 3.61$, B versus A, r:AB AB..., $n_{\text{prev}} = 100$, $n_{\max} = 200$, $x_0 = 0.5$. L-shading. LL:$(-1, -1)$, UL:$(-1, 4)$, LR:$(4, -1)$

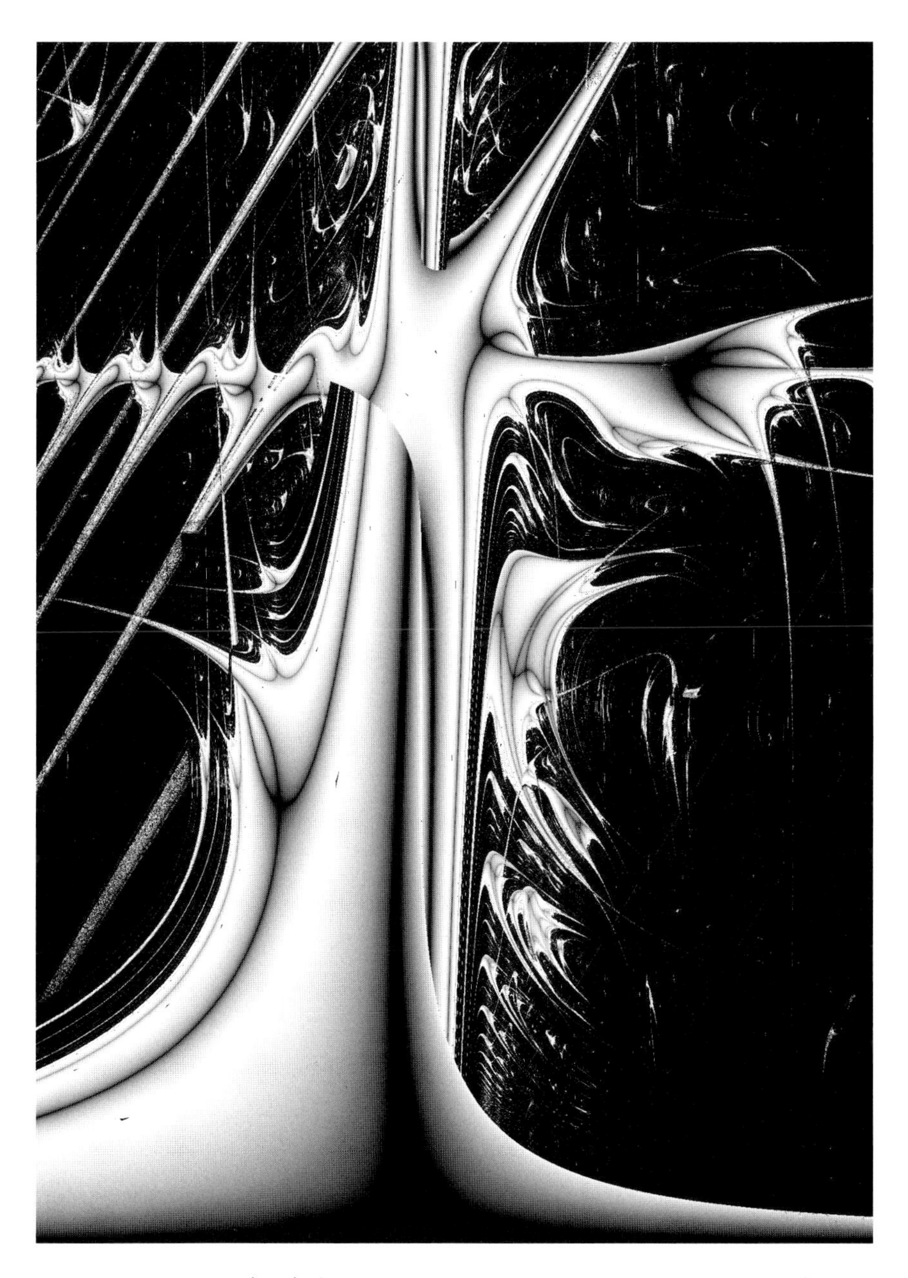

Fig. 9.144 Equation (9.76): b versus r. $n_{\text{prev}} = 100$, $n_{\max} = 200$, $x_0 = 0.5$. L-shading. LL:$(-3.2, 0)$, UL:$(-3.2, 3)$, LR:$(3.75, 0)$

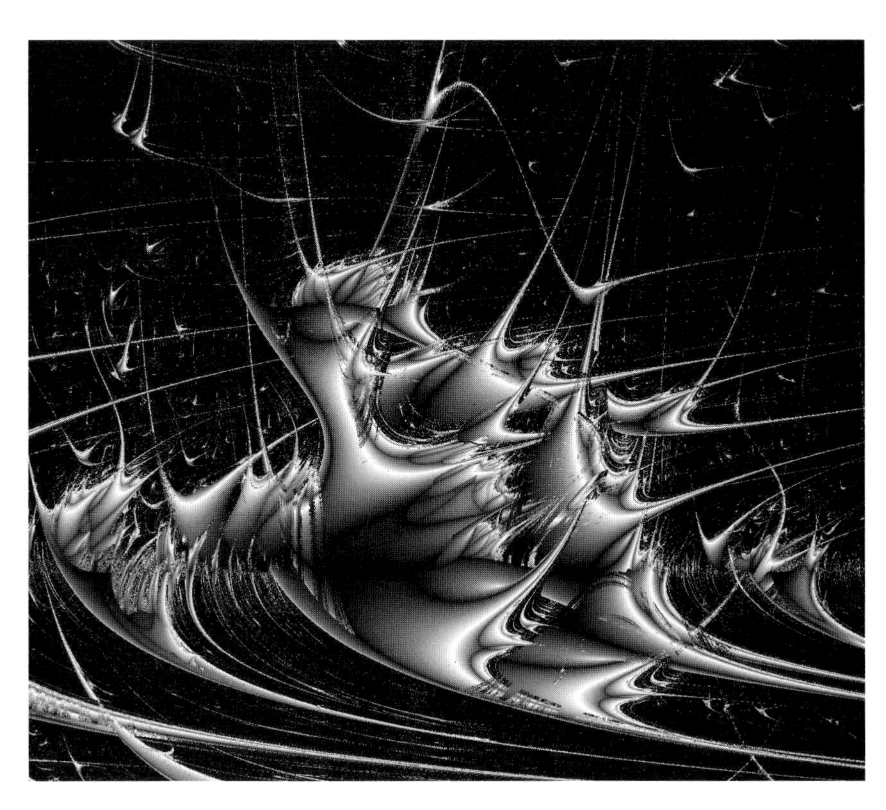

Fig. 9.145 Equation (9.77): $b = 0.9$, B versus A, r:AABB AABB..., $n_{\text{prev}} = 600$, $n_{\max} = 300$, $x_0 = 0.7$. D-shading. LL:(4.0915, 4.0223), UL:(4.0915, 4.0989), LR:(4.1804, 4.0223)

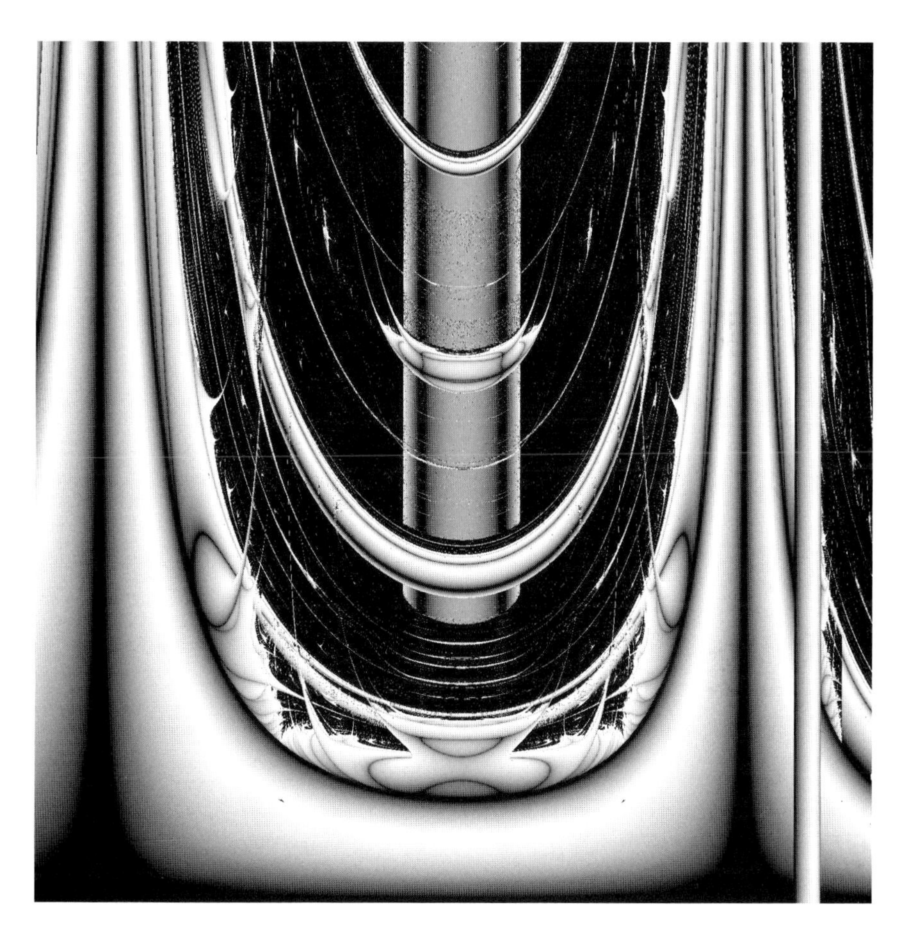

Fig. 9.146 Equation (9.77): $b = 0$, b versus r, $n_{\text{prev}} = 100$, $n_{\text{max}} = 200$, $x_0 = 0.5$. L-shading. LL:(0.9, 0), UL:(0.9, 2.7), LR:(2.2, 0)

Fig. 9.147 Equation (9.78): $b = 0.5$, B versus A, r:AB AB..., $n_{\text{prev}} = 500$, $n_{\max} = 1000$, $x_0 = 0.5$. D-shading. LL:(0, 0), UL:(0, 2), LR:(2, 0)

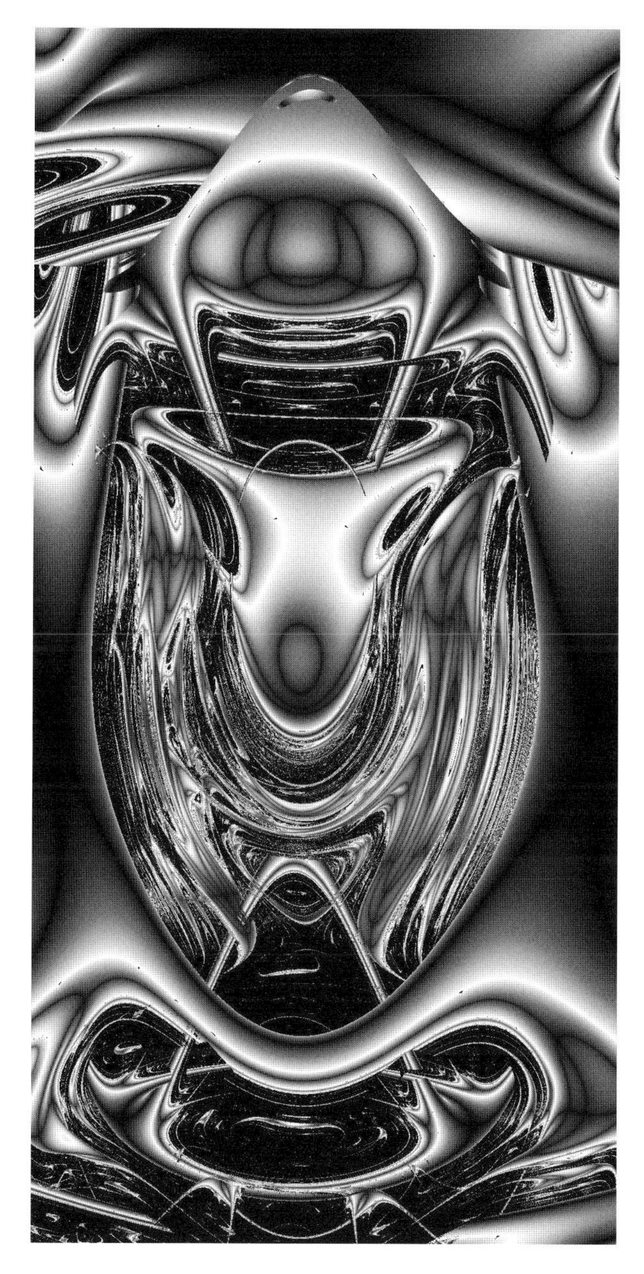

Fig. 9.148 Equation (9.79): $b = 1$, B versus A, r:AB AB..., $n_{\text{prev}} = 100$, $n_{\max} = 200$, $x_0 = 1.5$. D-shading. LL:(2.7741, 3.1152), UL:(2.7785, 2.6996), LR:(2.5144, 3.1125)

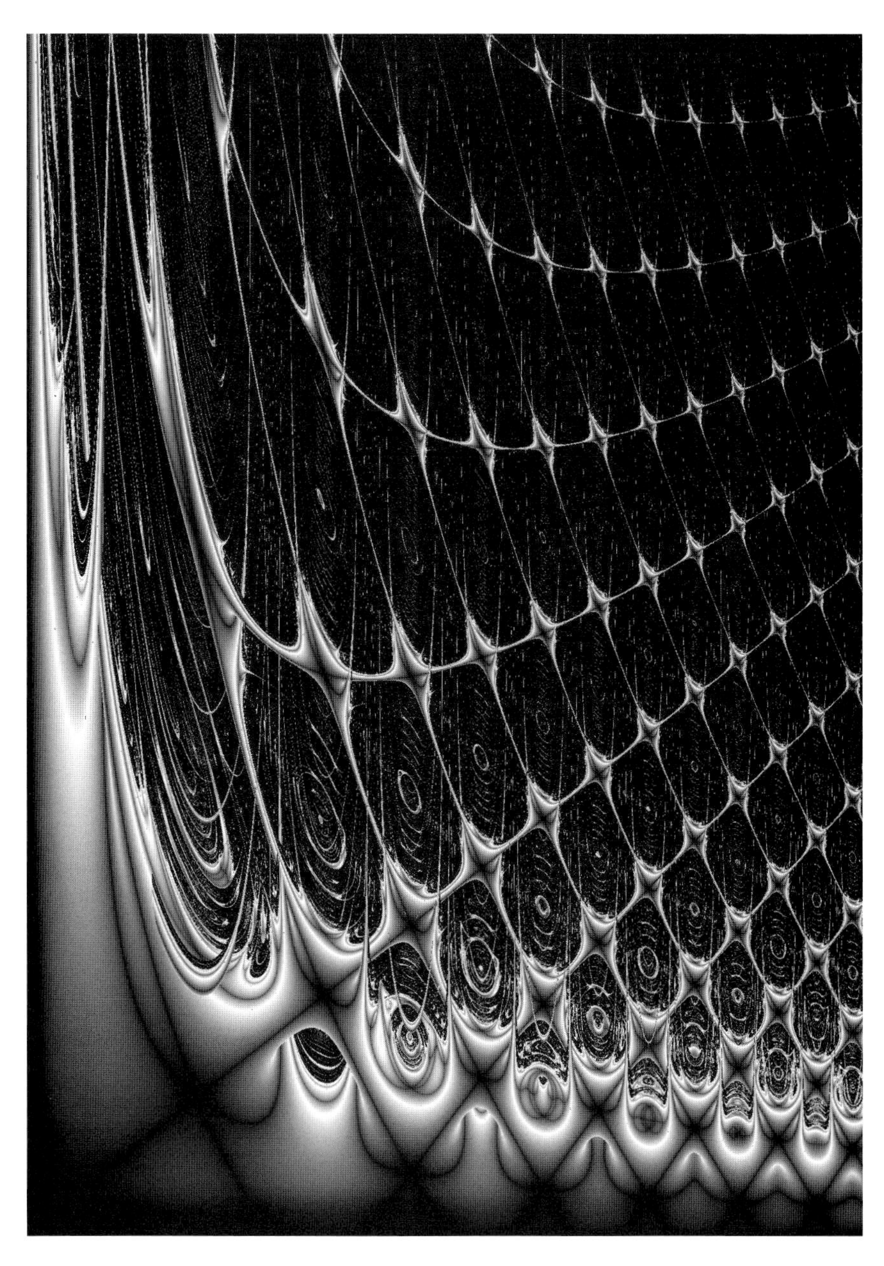

Fig. 9.149 Equation (9.79): $b = 0$, b versus r, $n_{\text{prev}} = 100$, $n_{\text{max}} = 200$, $x_0 = 0.5$. D-shading. LL:(0, 0), UL:(0, 7.66), LR:(4.3, 0)

Fig. 9.150 Equation (9.80): $b = 0.9$, B versus A, r:AB AB..., $n_{\text{prev}} = 50$, $n_{\max} = 100$, $x_0 = 1.4$. D-shading. LL:$(-5.5388, 2.20)$, UL:$(-5.5388, 3.6217)$, LR:$(-4.94297, 2.20)$

Fig. 9.151 Equation (9.80): $b = 0$, r versus b, $n_{\text{prev}} = 100$, $n_{\text{max}} = 200$, $x_0 = 0.5$. L-shading. LL:(0.26, 1.36), UL:(0.26, 3.85), LR:(1.44, 1.36)

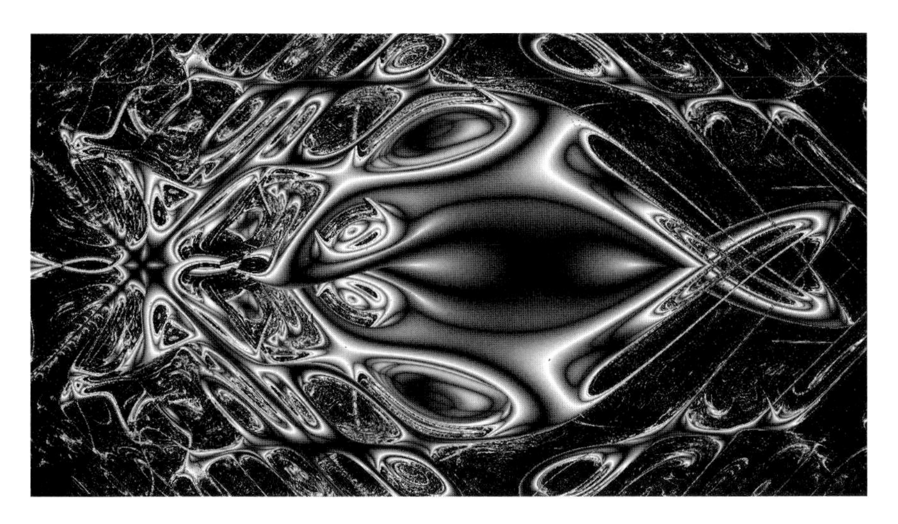

Fig. 9.152 Equation (9.59): $b = 0.6$, B versus A, r:AB AB..., $n_{\text{prev}} = 100$, $n_{\max} = 200$, $x_0 = 0.5$. D-shading. LL:(−0.025, −2.3), UL:(−2.3, −0.025), LR:(4.2, 1.925)

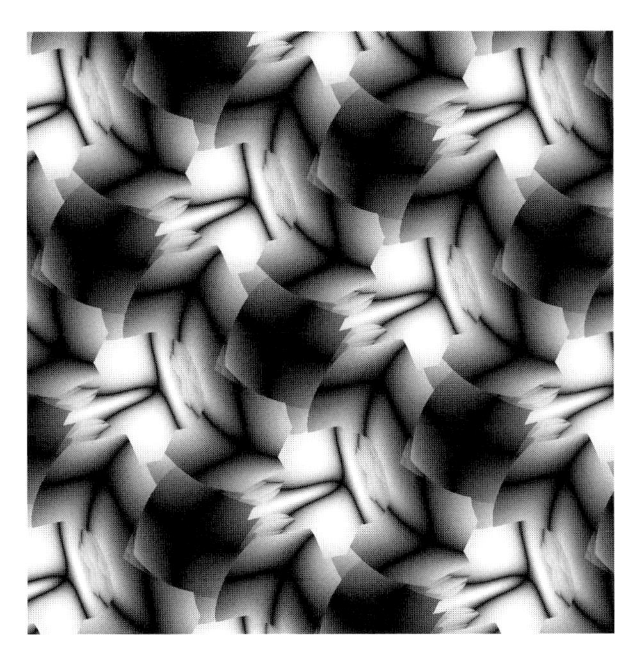

Fig. 9.153 Equation (9.81): $b = 2$, B versus A, r:AB AB..., $n_{\text{prev}} = 200$, $n_{\max} = 200$, $x_0 = 0.5$. D-shading. LL:(−4, −12), UL:(−12, 4), LR:(12, −4)

Fig. 9.154 Equation (9.81): $b = 0$, r versus b, $n_{\text{prev}} = 100$, $n_{\text{max}} = 200$, $x_0 = 0.5$. L-shading. LL:(−9.7, 0), UL:(−9.7, 9.9), LR:(7.95, 0)

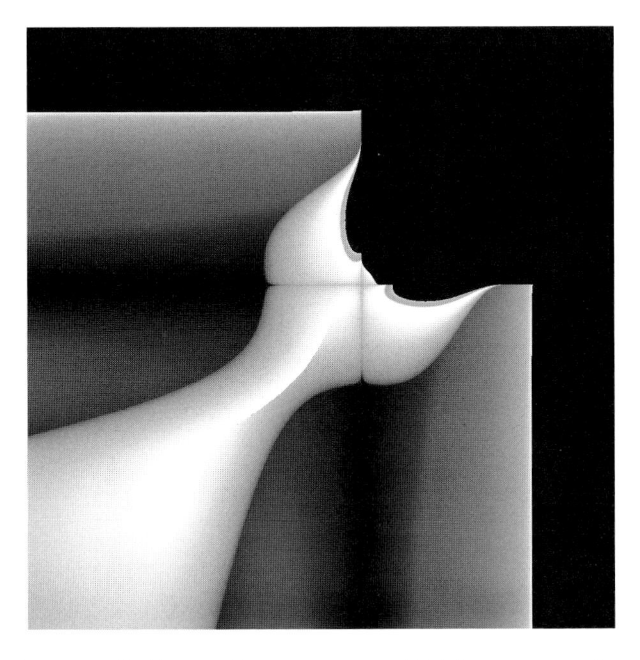

Fig. 9.155 Equation (9.82): $b = 0.2$, B versus A, r:AB AB..., $n_{\text{prev}} = 100$, $n_{\text{max}} = 200$, $x_0 = 0.5$. D-shading. LL:(−4, −4), UL:(−4, 3), LR:(3, −4)

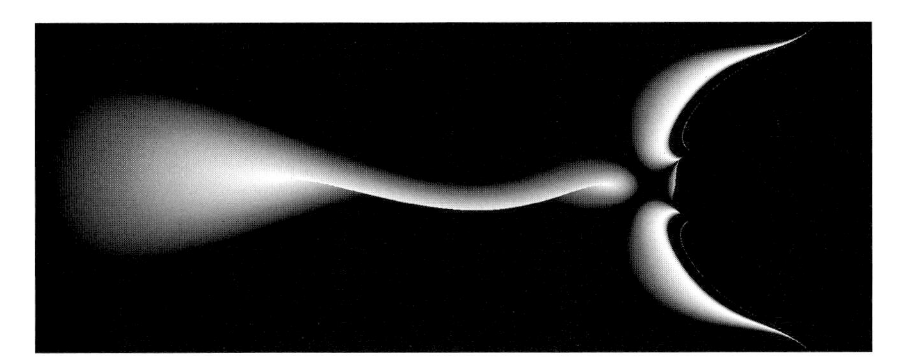

Fig. 9.156 Equation (9.82): $b = 1$, B versus A, r:AB AB..., $n_{\text{prev}} = 100$, $n_{\text{max}} = 200$, $x_0 = 0.5$. L-shading. LL:(−0.4615, −0.8326), UL:(−0.8073, −0.5013), LR:(0.4008, 0.0676)

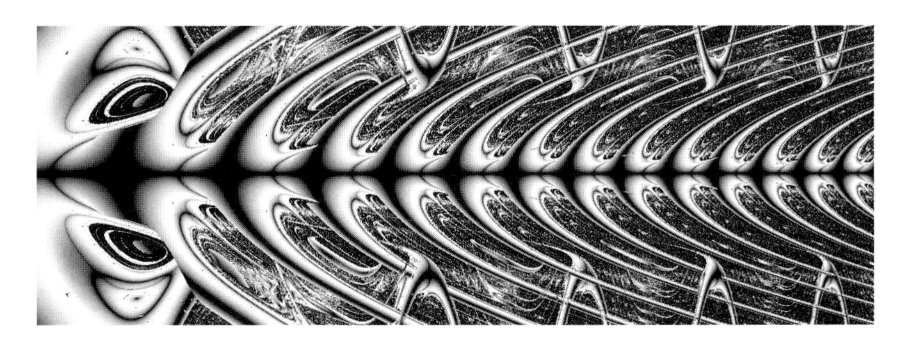

Fig. 9.157 Equation (9.83): $b = 0.8$, B versus A, r:AB AB..., $n_{\text{prev}} = 50$, $n_{\text{max}} = 100$, $x_0 = 0.5$. L-shading($\lambda_m = -2$). LL:(1.1481, −0.6903), UL:(1.1481, 0.6741), LR:(5.0651, −0.6903)

Fig. 9.158 Equation (9.59): $b = 0.6$, B versus A, r:AB AB..., $n_{\text{prev}} = 25$, $n_{\max} = 50$, $x_0 = 0.5$. L-shading. LL:$(-0.988, 0.421)$, UL:$(1.4136, 1.5955)$, LR:$(-0.0746, -1.4467)$

Fig. 9.159 Equation (9.84): $b = 0.3$, B versus A, r:AB AB..., $n_{\rm prev} = 500$, $n_{\rm max} = 1000$, $x_0 = 0.5$. L-shading($\lambda_m = -2$). LL:(-6, 0), UL:(0, 6), LR:(0, -6)

Fig. 9.160 Equation (9.85): $b = 0.5$, B versus A, r:AB AB..., $n_{\text{prev}} = 100$, $n_{\max} = 200$, $x_0 = 0.5$. L-shading($\lambda_m = -1$). LL:(2.0292, 2.3214), UL:(2.5354, 2.832), LR:(2.3177, 2.0354)

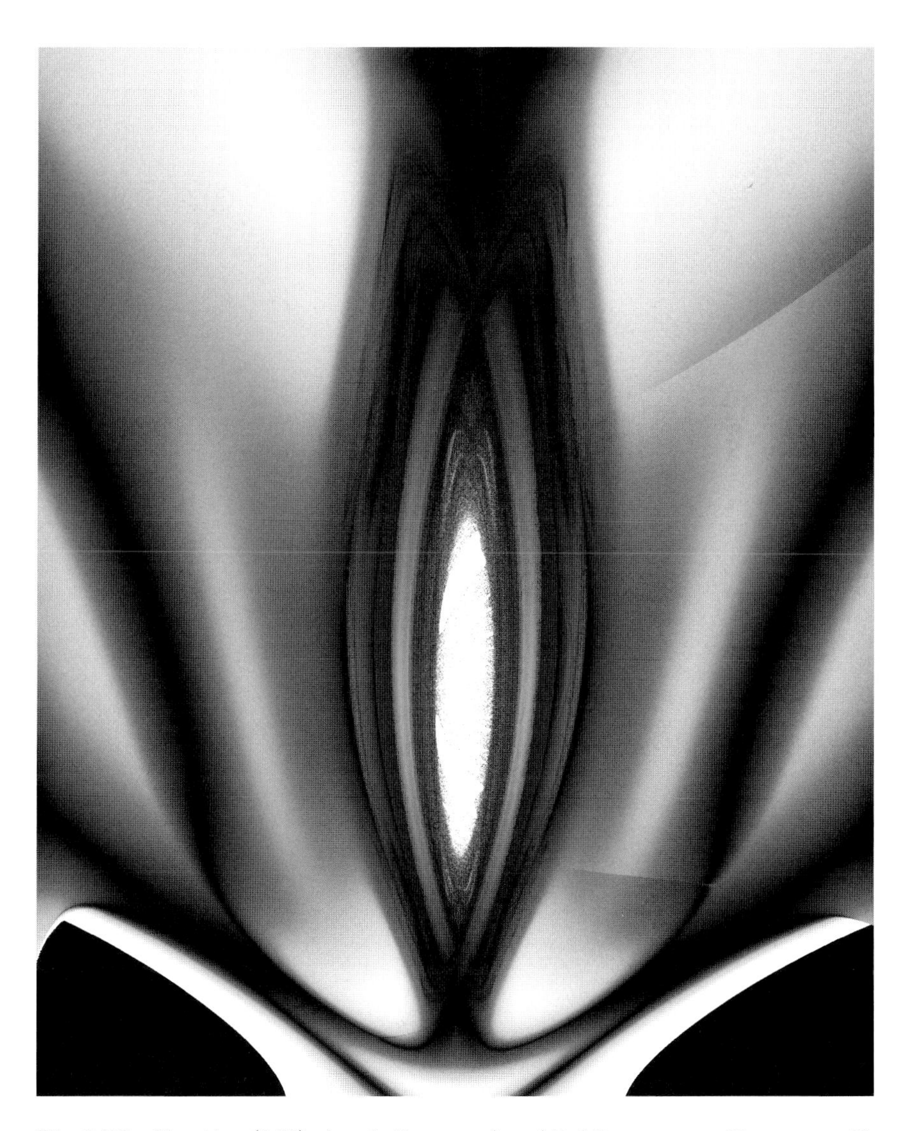

Fig. 9.161 Equation (9.86): $b = 1$, B versus A, r:AB AB..., $n_{\text{prev}} = 25$, $n_{\text{max}} = 50$, $x_0 = 0.5$. D-shading. LL:$(-0.2134, 0.2861)$, UL:$(0.5603, 1.0899)$, LR:$(0.296, -0.2042)$

Chapter 10

Are the λ-Diagrams Fractals?

Some readers may answer "yes" to the question posed in the chapter title, as it has become a habit to call any complicated pictures arising from the context of chaos "fractals". However, the correct answer is that most λ-diagrams are not fractals, while some of these diagrams are so-called "fat fractals".

In order to understand this answer, let us first formally define a fractal on a plane. For this we cover the 2D object with squares with a side length of ϵ. Let $N(\epsilon)$ be the number of squares containing at least one point of the object. Then, $N(\epsilon) \sim \epsilon^{-F}$ for $\epsilon \to 0$, where F is the fractal dimension.

The measure of the object is defined as $\mu(\epsilon) = N(\epsilon)\epsilon^2$ for $\epsilon \to 0$. Thus $\mu(\epsilon) \sim \epsilon^{2-F}$. For fractals such as the Hénon attractor shown in Fig. 7.12, F is a non-integer between 1 and 2. Note that this non-integer dimensionality is a feature of all so-called "thin fractals". In fact, a coastline made up of bays, which themselves contain smaller bays, and so on, has a fractal dimension between 1 and 2, as it is something between a curve and a plane. A tree or our blood vessels have a fractal dimension between 2 and 3 because they are less space-filling than 3D space, but are more space-filling than a surface. Returning to the measure $\mu(\epsilon)$ in two dimensions, $F < 2$ implies that $\mu(\epsilon) \sim \epsilon^{2-F} \to 0$ for $\epsilon \to 0$.

The fact that $N(\epsilon) \sim \epsilon^{-F}$ is independent of ϵ reflects the fact that we are dealing with a self-similar object, i.e. an object that looks similar at any size-scale. We see this feature in Fig. 7.12, as well as in coastlines, bloodvessels, cauliflowers and other well-known fractals. If one observes that some λ-diagrams are also self-similar (for example, Plate 10 is an enlargement of a small subset of Plate 9; see also e.g. Plates 13 and 15) then one could claim that we are dealing with fractals. However, one clearly sees that our objects have well-defined surfaces inside, so that they have by no means zero measure, as was deduced above. How do we solve this riddle? The answer is quite simple: $N(\epsilon) \sim \epsilon^{-F}$ with $F = 2$ and thus $\mu(\epsilon) \sim \epsilon^{2-F} = \epsilon^0$, which does not go to zero as $\epsilon \to 0$. Objects having this property are called "fat fractals" [126, 127], in contrast to the fractals described above, which are consequently called "thin fractals". Examples of fat fractals are the human body minus all blood vessels, as well as the well-known Mandelbrot set (Fig. 9.25; Section 9.18).

The problem arising for fat fractals is the missing non-integer dimension that so elegantly characterizes thin fractals. In fact, $F = 2$ for all fat fractals on the plane and $F = 3$ for all fat fractals in three dimensions. As a way out, it was found that fat fractals can be characterized by the following scaling law of their measure $\mu(\epsilon) = N(\epsilon)\epsilon^D$, where D is the dimension of the space where the fractal is embedded (D is equal to 2, 3, etc.):

$$\mu(\epsilon) = \mu_0 + C\epsilon^\gamma \ . \tag{10.1}$$

The so-called fatness exponent γ is used to characterize a fat fractal. γ was determined for some λ-diagrams, the object covered with squares being the set defined by $\lambda < 0$.

In previous investigations of fat fractals, the paramters μ_0, C and γ were determined by two methods: (a) Determination of μ_0 by extrapolation of μ for $\epsilon \to 0$ and subsequent plot of $\log(\mu - \mu_0)$ versus $\log(\epsilon)$; and (b) non-linear optimization. In general, method (a) yields highly uncertain estimates of μ_0, unless the values of ϵ are so small that computing capacity is at the edge of today's possibilities. On the other hand, method (b) has the drawback that more than one local minimum of the sum of squares SSQ (sum of squares of differences between data and values obtained from the

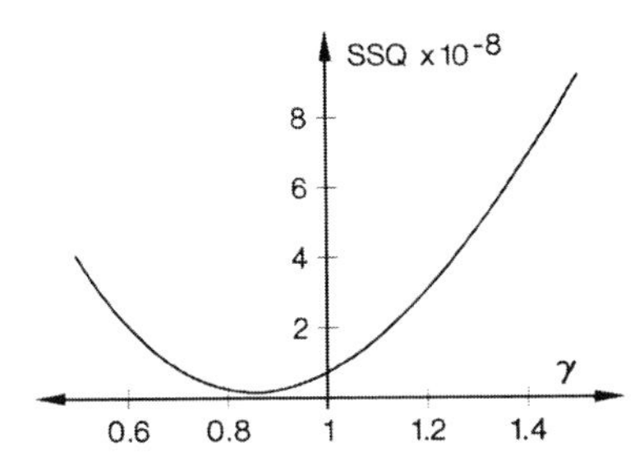

Fig. 10.1 Minimum of the sum of squares (SSQ) versus γ after a linear fit of μ versus ϵ^γ . (Computed for Fig. 10.3). The minimum of the SSQ determines the optimal γ.

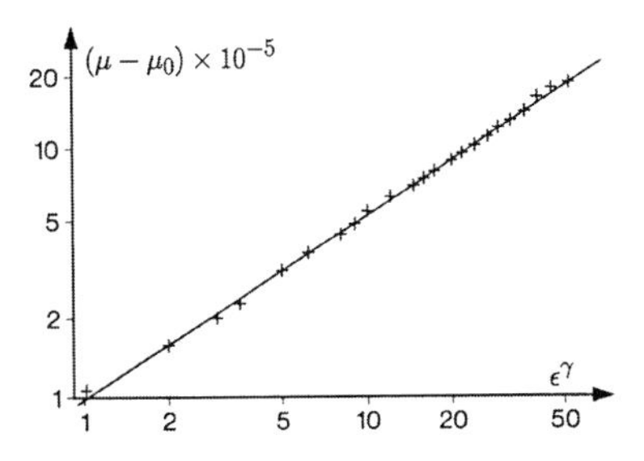

Fig. 10.2 Log-log plot of $\mu - \mu_0$ versus ϵ^γ, computed for the value of γ at which the SSQ is minimum in Fig. 10.1.

Fig. 10.3 Fat fractal corresponding to Plate 10; black: $\lambda < 0$; white: $\lambda > 0$.

equation) may be obtained. The problem of finding the global minimum is, indeed, common to many non-linear optimization problems. In Ref. [127], this problem was solved by the following method.

For a given γ, μ versus ϵ^γ is fitted by linear optimization. Then γ is changed in small steps and the SSQ obtained from each linear optimization is plotted versus γ. Then, the γ for which the SSQ is mimimum is determined. This is exemplified in Fig. 10.1. The plot corresponds to the set with $\lambda < 0$ in Plate 10, which is black in Fig. 10.3.

The values of γ given below were calculated with a resolution of 2160×2160 points. The number 2160 has the advantage of being divisible by a large number of integers, all of which are used as ϵs: 1, 2, 3, 4, 5, 6, 8, 9, 10, 12, 15, 16, 20, 24, 27, 30, 36, 40, 45, 48, 54 and 60. Because of the divisibility by these numbers, no residuals were left when covering the graphs with these ϵs.

The error of γ was determined (with a confidence of 95%) by considering it composed of two contributions: a deviation appearing in the linear regression, and a deviation stemming from the curvature (see Fig. 10.1) around the minimum of the SSQ versus γ. The first was obtained as a standard by-product of linear regression. The latter was determined using the F-test, which yields a quotient of SSQs (ordinate) corresponding to an error-interval of γ (abscissa); according to the F-test, γ-values within this interval have no significantly different variances, i.e. different SSQs, at the given confidence of 95%.

Here are some exemplary results for γ:
Plate 9: 0.837 ± 0.13. Plate 10: 0.853 ± 0.128. Plate 15: 0.671 ± 0.071. Plate 17: 0.544 ± 0.055. Plate 19: 0.451 ± 0.063. Plate 20: 0.859 ± 0.161. Plate 21: 0.401 ± 0.068. Plate 22: 0.345 ± 0.085. Plate 23: 0.885 ± 0.148. Plate 24: 0.612 ± 0.061. Plate 25: 0.572 ± 0.051. Plate 27: 0.823 ± 0.064. Plate 32: 0.738 ± 0.059.

One may ask now what is gained by knowing these numbers. An answer is that they yield an excellent linearity for $\mu(\epsilon) - \mu_0$ versus ϵ^γ, as exemplified in Fig. 10.2. This linearity demonstrates that the λ-diagrams have a size-scale independent property, i.e. that they display a quantitative self-similarity. Although this self-similarity is not describable by a fractal dimension, it is describable by a well defined quantity, namely the fatness exponent γ.

For many diagrams in this book, e.g. Figs. 8.4, 8.6, 9.137, 9.142 and 9.156, there is no self-similarity, as can be seen by mere visual inspection. Thus, these diagrams are not fractals of any kind.

Chapter 11

What Can We Learn from λ-Diagrams?

Apart from the aesthetical value that the diagrams may have, they are a scientific tool. This tool allows to have a rapid visual overview of the stability of a system under variation of control parameters (coordinates). The idea is that, instead of looking for a certain property with a search algorithm, one scans the parameter plane in an indiscriminate way, and then just throws a glance on it. The disadvantage is the length of computing times, while the advantage is the appearance of unforeseen features. Let me express this advantage as a metaphor: in place of sailing, like Columbus, one explores the world via satellite. Particular phenomena that can be detected by a glimpse on the plane are:

(a) Transitions between predictability ($\lambda < 0$) and chaos ($\lambda > 0$). These transitions are usually indicated in the diagrams by a discontinuity in the colour or in the shade of grey.

(b) Different stability of predictable behaviour ($\lambda < 0$), as indicated by changing colours or shades of grey. In the latter case, the shading changes (in this book) from black to white, as λ changes from its minimum (negative) value to zero. $\lambda < 0$ implies periodicity and perturbations recover proportionally to $e^{-|\lambda|t}$. Thus, $|\lambda|$ for $\lambda < 0$ tells us how fast a periodic system recovers after a perturbation.

(c) Different degrees of chaos. This is also indicated by varying colours or shades of grey in regions where $\lambda > 0$. When shades of grey are used, these regions are usually coloured black in this book, but one can as well distinguish different ranges of positive λ, as it was done here in many colour plates. Perturbations grow proportionally to $e^{|\lambda|t}$. Thus, for a small $|\lambda|$ predictions are possible for a longer time in advance than for a larger $|\lambda|$.

Given an initial error of the measurement or the calculations, the product of this error and $e^{|\lambda|t}$ can be set equal to the size of the system. Solving the resulting equation for t tells us the time within which the initial error dominates the whole system, thus making any prediction impossible.

(d) Points or curves in which periodical behaviour is extremely stable, i.e. λ is extremely negative, including the particular case $\lambda \to -\infty$, i.e. the so-called superstability. In the black and white diagrams of this book, they usually appear as darker curves within the (lighter) regions where $\lambda < 0$. In the life sciences, for example, it is interesting to know how close evolution has driven biorhythms to such a condition. In technology, this feature is interesting in the optimization of robust devices.

(e) Coexistence of attractors. Whenever branches overlap, as illustrated in Fig. 11.1, there exists more than one attractor for the same parameter values. When browsing through this book one finds a large number of such overlaps. In such cases, the attractor that is reached by the system depends on the initial conditions and corresponds to the branch that covers the other in a diagram. For different initial conditions than those used in the figure, the upper branch may be below. In addition, overlapping branches indicate hysteresis: if parameters are (slowly) changed then one or the other attractor will be reached, depending on the previous values of the parameters, i.e. on the system's history. Browsing through this book, one may find three or more branches overlapping in the same region of the plane; this means that there exist the corresponding number of coexisting attractors (three

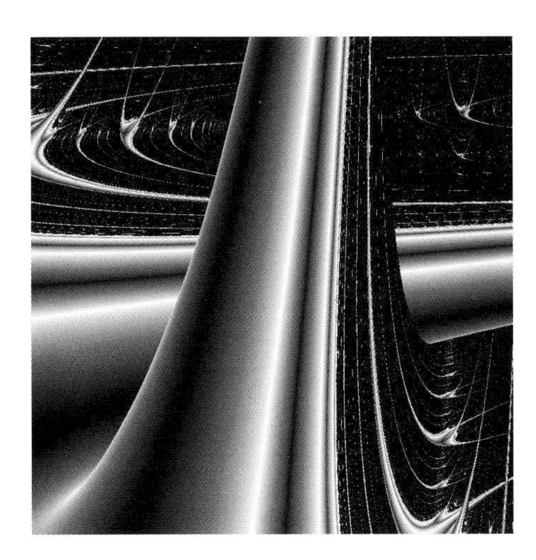

Fig. 11.1 Example of overlapping branches, indicating the coexistence of attractors. Equation 9.4 B versus A. r:AB AB..., $x_0 = 0.51$, $n_{prev} = 5000$, $n_{max} = 5000$, D-shading. LL:(2.9, 2.9), UL:(2.9, 4.), LR:(3.94, 2.9)

or more), and thus multiple hystereses such as those discussed in Ref. [19].

One of the most eye-catching features of many diagrams on the A-B-plane is that symmetry with respect to the line A=B is not perfect. (Note: the line A=B is the diagonal from the lower left to the upper right if coordinates are not rotated; often, however, A=B corresponds to a vertical in the middle, due to a rotation of coordinates by 45°). The lack of perfect symmetry with respect to the line A=B is mainly due to the coexistence of attractors. In fact, since the A-B-sequence starts with A or with B, the parameters A and B are not interchangeable. Thus, one coexisting attractor may be reached at the point (A, B) and a different attractor at the point (B, A), entailing the asymmetry of the diagrams. This "rupture of symmetry" is considered by many observers as an aesthetically appealing feature of the diagrams.

(f) Coexistence of attractors with fractal or riddled basins. Whenever overlapping branches look "transparent" (as in the example in Fig. 11.2) the attractors corresponding to these branches have strongly interleaved, fractal or riddled basins (see [8, 128]) in phase space. (See also Sections 9.13 and 9.14). This is explained as follows.

On the plane, the initial values x_0 or (x_0, y_0) are fixed. As one moves slightly on such a "transparent" region of the plane, the parameters are shifted and so are the highly interwoven basins. This causes the (fixed) initial values to be alternately on the basin of one attractor and on the basin of the other. Thus, the two branches alternate very closely together in the diagram, which gives the impression of transparency. Note that in

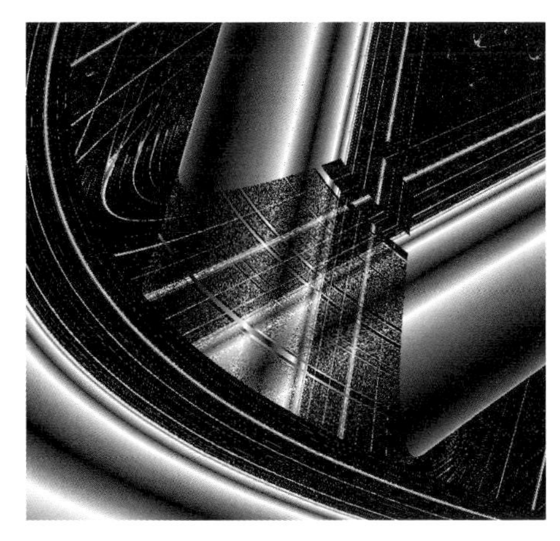

Fig. 11.2 Example of "transparent" looking, overlapping branches (compare with Fig. 11.1), indicating strongly interleaved basins of attraction. These basins turn out to be fat fractals on further analysis [8]. Equation (9.4): B versus A. r: BBABABA BBABABA, $x_0 = 0.5$, $n_{prev} = 10000$, $n_{max} = 10000$, D-shading. LL:(3.8358, 3.5952), UL:(3.8358, 3.6057), LR: (3.8481, 3.5952).

some cases one does not get the visual impression of "transparency" in the case of highly interwoven basins; examples are given in Figs. 9.19 and 9.20 (Section 9.14), which indicate basins riddled to each other, i.e. intermingled basins.

Fractal basins were analysed in detail for the map defined by Eqs. (9.27, 9.28) in section 9.13. For that map, the fractal dimension of the boundary between the two basins is $d \approx 1.8$. If one desires to reach a given fixed point from a given initial condition, and the error of this initial condition is ϵ, then one would reach the wrong fixed point in a fraction

$$f \sim \epsilon^{D-d} \tag{11.1}$$

of all runs [111]. D is the dimension of the phase space, i.e. $D = 2$ in this case. Thus, if for an error $\epsilon = \frac{1}{8}$, 66% of the initial conditions lead to the wrong result, then the much smaller error $\epsilon = 3 \cdot 10^{-5}$ reduces the percentage of failure only to 12%. In other words, a small uncertainty in the initial conditions yields a substantial fraction of calculations which are uncertain as to which final state is eventually attained. This is certainly a different unpredictability than that related to the "butterfly effect", in which small perturbations of the initial conditions grow in time for a chaotic system. Contrarily to chaos, where predictions for sufficiently small times are possible, the time between prediction and predicted event plays no role in the present case: the situation is like the flipping of a coin. The situation is especially dramatic if the basins are riddled or intermingled (Section 9.14; Figs. 9.19 and 9.20), since then the "uncertainity exponent" $D-d$ is nearly zero; in that case a reduction in the error ϵ has no significant effect on the fraction f of wrong asymptotic behaviour.

(g) Strongly interleaved zones of changing sign of λ, implying structural instability, i.e. a sensitive dependence of the dynamics (predictable or chaotic) on the control parameters (See Chapter 5). Examples can be found in Plates 13 and 30 and in many other diagrams throughout this book. (Note: fractal or riddled basins imply structural instability, but not necessarily the other way around.) And a word of caution: a definitive conclusion about structural instability should only be made after showing that the changes in the sign of λ remain if one increases the number of iterations. In fact, if the number of iterations is too low, the evaluation of λ may be dominated by transients, instead of attractors, and transients often display (transient) interleaved regions.

(h) Trajectories on or close to a torus, i.e. quasiperiodical or nearly quasiperiodical dynamics. This occurs, for example for strange nonchaotic attractors (see Section 9.19). λ-diagrams then display large areas with different shades of grey or colours that keep changing shape as iterations go on. These large areas indicate fluctuations close to the value $\lambda = 0$. Note that if the dynamics take place exactly on a torus, the λ-diagrams do not converge for $n \to \infty$.

Acknowledgments

I would like to thank the following people for their assistance during the writing of this book: My doctorate student Malte Schmick, who collaborated with me in the making of the CD-ROM; Andreas Ehrhard and Susanne Meier, who provided assistance in the production of colour pictures; and my student Alexander Hasselhuhn, who helped with the formatting and layout of the text and figures.

In addition, I shall list former students who collaborated with me in the making of λ-diagrams, very much in the manner of those who delight in cooking together. As the reader may discover, teamwork is indeed useful in the often long-winded search, selection, scaling, spatial orientation and colouring of the diagrams. Fifteen per cent of the λ-diagrams in this book resulted from my interaction with these students. I thank them for the fun we had together!

Karsten Kötter: Figs. 9.124, 9.145, 9.147
Mathias Woltering: Figs. 9.120, 9.132, 9.133, 9.135, 9.161
Katrin Sulzbacher: Figs. 9.116, 9.129, 9.130, 9.137, 9.139, 9.140, 9.141 and 9.157
Andreas Gasper: Plates 2, 3, 29, 30, 31, 33, 34, 37, 38 and 42
Hans Schepers: Plate 1
Martin Allin: Plates 25, 27, 28 and 32
Tobias Kauch: Figs. 9.136, 9.160
Jan-Martin Wischermann: Fig. 9.142
Christian Krüger: Plates 41 and 43

I thank Jaime Rössler, outstanding scientist and friend, for a crucial remark over lunch: "Hey, Mario, why not assume that a parameter varies, say like ABAB..., and look at the dynamics on the A-B-plane?". It is incredible that a simple sentence like that could cause such an avalanche of graphs!

One person deserves more than a few words: Bárbara Sölter who has been encouraging me for decades and organized an exhibition in Chile which won a prize for "Exhibition of the Year".

Also I would like to thank my former boss, Benno Hess (in memoriam), who recognized that these were more than "pretty pictures", bought for me a colour monitor (1000×1000 pixels, and at a time when it cost 80 thousand dollars) and organized exhibitions all over the world.

Finally I thank Katie Lydon and Lizzie Bennett for their help with the English wording, Jimmy Low Chye Chim for the design of the book's cover, Linda Kwan for managing the production of the book, Lance Sucharov for opening the door to Imperial College Press to me, and Luis Cortés for the portrait of me in the jacket sleeve.

Appendix A

Informal Glossary

λ-diagram: Term used in this book for the representation of the Lyapunov exponent on a plane defined by two control parameters.

Attractor: Trajectory in phase space for a dissipative system; it is reached after starting from nearby initial conditions.

Basin of attraction: Set of initial conditions leading to a given attractor.

Chaos: Situation in which the maximum Lyapunov exponent is positive. For a given time in advance, only predictions with limited accuracy are possible. Usually, its occurrence corresponds to the backgrounds of the λ-diagrams.

Coexistence of attractors: Existence of more than one attractor, each one with its own basin of attraction in phase space.

Conservative system: System in which energy is conserved.

Control parameter: Quantity that is held constant in time.

Dissipative system: System in which energy is dissipated.

Fat fractal: Geometrical object that is self-similar at all length scales, but is described by an integer dimension. It can be characterized by a so-called fatness exponent.

Fractal: Geometrical object that is self-similar at all length scales and is describable by a non-integer dimension.

Intermingled basins: Basins that are riddled to each other.

Lyapunov exponent: Exponent describing the mean exponential growth or decay of a small perturbation of a trajectory.

Parameter space: Space defined by control parameters.

Periodicity: Dynamics described by a closed trajectory in phase space. The maximum Lyapunov exponent is negative. Predictions are possible. Usually, it occurs in the "figures" at the foregrounds of the λ-diagrams.

Phase space: Space defined by the phase variables.

Phase variable: Quantity that can change in time.

Poincaré section: Cut with a plane through phase space.

Quasiperiodicity: Dynamics on a torus in phase space. It is describable by two or more incommensurate frequencies. The maximum Lyapunov exponent is zero.

Riddled basins: The basin in phase space of an attractor A is riddled to that of an attractor B if in any neighbourhood of an initial condition leading to A there exist initial conditions leading to B.

Strange attractor: Fractal chaotic attractor.

Structural instability: Situation with highly interleaved regions (corresponding to different modes) in parameter space.

Superstability: Case in which the Lyapunov exponent $\to -\infty$

Transient: Trajectory on the way to an attractor from a point on its basin of attraction.

Appendix B

Abbreviations

λ: Maximum Lyapunov exponent. $\lambda < 0$ indicates periodicity, i.e. predictability (usually in the foreground of the diagrams). $\lambda > 0$ indicates chaos, i.e. unpredictability (usually in the background of the diagrams and usually black if shades of grey are used).

λ_m: minimum negative λ for L-shading imposed by the user.

BZ: Belousov-Zhabotinsky reaction

D-shading: "Democratic shading". Equal number of pixels with negative λ are assigned to each grey level 1 through 255.

LL: coordinates of the lower left corner of a diagram.

LR: coordinates of the lower right corner of a diagram.

L-shading: "Linear shading". The interval between the minimum and the maximum negative λ is divided into 255 subintervals. The grey levels 1 through 255 are assigned to these subintervals.

$\mathbf{n_{max}}$: number of iterations for the computation of λ; these iterations are started after finishing n_{prev} iterations.

$\mathbf{n_{prev}}$: number of iterations for transients to die away; these iterations are performed before the computation of λ is started.

r: AAB AAB ... (example): The parameter r assumes the value A, then A, then B, then again A, A and B, and so on periodically, as the iterations proceed.

SSQ: sum of squares of differences between the values of the data and of a model equation

UL: coordinates of the upper left corner of a diagram.

$\mathbf{x}$, $\mathbf{y}$, $\mathbf{z}$: phase variables

$\mathbf{x_0}$, $\mathbf{y_0}$ **and** $\mathbf{z_0}$: Initial values of the phase variables.

$\mathbf{x_n}$, $\mathbf{y_n}$ **and** $\mathbf{z_n}$: n-th iterates of the phase variables.

Appendix C

Instructions for the CD-Rom (λ-Diagrams on Your PC)

Contents:

C.1 General information

Those who understand the maths in Chapter 6 and who are versed in an arbitrary computer programming language should have no problems in reproducing or creating λ-diagrams on their PC without the CD-ROM. Independence from the CD-ROM will give the reader more flexibility and control, but will be more demanding.

A more comfortable approach is provided by the CD-ROM that accompanies this book. This CD-ROM enables λ-diagrams to be created with the included program "Lyap.jar", which is written in **Java** and is **compatible with Windows, MacOS, Linux/Unix**[129]. To run it, your computer must have the Java Runtime Environment installed. Installation files for Windows and Linux are provided on the CD, as well as installers for the Java Development Kit, which is needed to provide your own equations for accelerated computations. For users of Windows, the CD has an auto-play function and the program should start automatically when the CD is inserted. The program is located in the folder "Markus_Diagrams" on the CD.

After opening this folder in your file-browser, double-click on "Lyap(.jar)" to start the program. If this does not work, open the above directory in a command console (e.g. "Start"+"Execute"+"cmd" for Windows as operating system) and type "java -jar Lyap.jar". More hints on trouble-shooting are given in the file "README.TXT" in the same directory.

Lyap.jar is a straightforward implementation of the algorithms described in Chapter 6. It iterates equations like those given in this book, evaluates the Lyapunov exponent λ and displays the values of λ as grey-shaded or coloured pixels. When the equations are entered, the corresponding derivatives will be calculated automatically, using an adapted version of the Java math tools published by a group in the Hobart and William Smith Colleges [130].

C.2 How to run the calculations and specify what exactly to calculate

To warm up and get a sense of how the program works, just press "start" in the bottom right of the main window (shown in Fig. C.1). A λ-diagram calculated from the data of Plate 10 (Equation 9.4), which is the default setting of the program, will appear. Pressing "start" a few more times will refine the picture until it resembles Plate 10. Then, in the "colour input window" (see Fig. C.5 and Section C.3), which is always open, you can move the sliders and press "redraw" (after each new setting of the sliders) to get a feel for the colouring procedure.

The warm-up exercise above is done under the default setting in which "use LIB equation No.:" is not checked. The option with LIB equations will be discussed in detail in Section C.6. At this point, however, some examples from the LIB equations will be presented, again as warm-up exercises. For this, first check "1D" or "2D". Then check "use LIB equation No". Then type the number 1, 2, 3 or 4 (four examples were incorporated for 1D and four examples for 2D) in the entry field to the right and press "ENTER" after that number. Then press "start" and (if necessary) "reset exponents" a few times. One may colour with the "colour input window", including the choice of "democratic" colouring. The following diagrams will be obtained for 1D: (1) Plate 10; (2) Fig. 9.101; (3) Fig. 9.69; and (4) Fig. 9.116. For 2D: (1) Fig. 8.19; (2) Fig. 8.24; (3) Fig. 9.14; and (4) Fig. 9.3. Now let

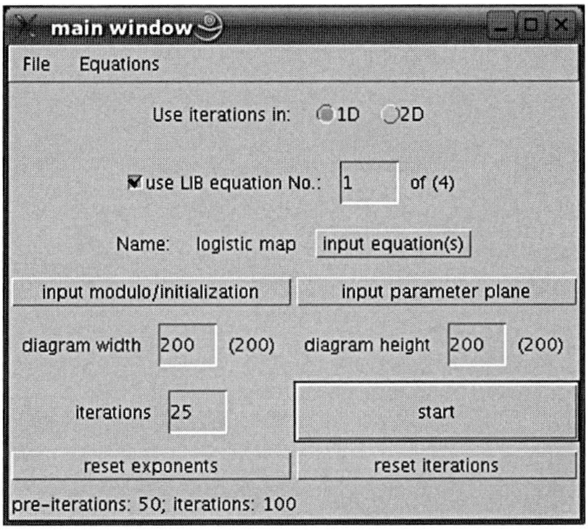

Fig. C.1 Main control window

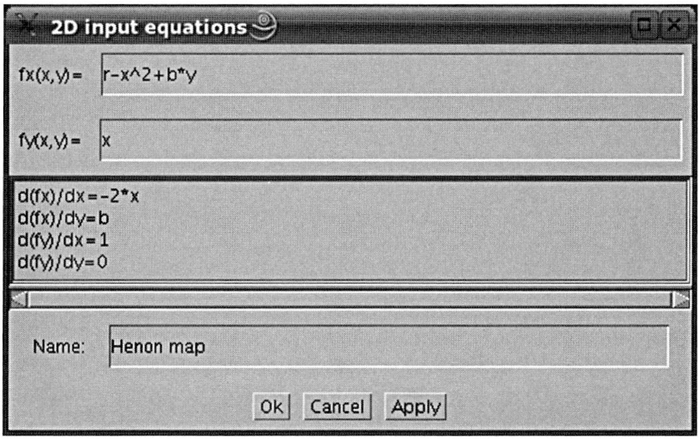

Fig. C.2 Window for "input equation(s)"

us forget the LIB equations, and delete the checking of "use LIB equation No".

To work creatively and seriously, inputs should be more individualistic than in the preceding prefabricated warm-up actions. Starting again from scratch, you must first decide if you want to work with 1D or 2D iterative equations (see Chapter 6) and mark your decision in the main control window (Fig. C.1). (The next option "use LIB equation No.:" is described below in Section C.6, and should not be chosen now.) Next, you put in an equation of your choice using the button "input equation". This opens a window like that in Fig. C.2. It is recommended that you begin with equations from this book and explore the "magic" of blow-ups and colouring; this could rapidly catapult the user beyond these book's images! Enter $f(x)$ for 1D, according to Equation (6.5), or the pair $\{f_x(x, y),\ f_y(x, y)\}$, according to Equations (6.6) and (6.7). The derivatives below are calculated automatically and updated, once "Apply" is pressed. The line marked "Name" may also be completed to show the inserted equation(s). The program recognizes the following input expressions:

- **x, y** – variables; **r** – parameter, which may be alternated as a sequence of values of A and B; **b** – parameter
- **e** – Euler constant 2.7182818...; **pi** – 3.1415927...
- **+,-,*,/** – elementary operations (plus, minus, times, divide by)
- **^** or ** – exponentiation
- **sqrt**(x) – square root of x; **cubert**(x) – cube root of x
- **abs**(x), **sin**(x), **cos**(x), **sec**(x), **csc**(x), **tan**(x), **cot**(x), **arcsin**(x), **arccos**(x), **arctan**(x), **exp**(x), **ln**(x), **log2**(x), **log10**(x)
- **trunc**(x) – drop the fractional part; **round**(x) – round to the nearest integer; **floor**(x) – round to the next lesser integer; **ceiling**(x) – round to the next greater integer
- =, <>, <, >, <=, >= – relations (equal, not equal, lesser, greater, lesser or equal, greater or equal)
- **()** – brackets
- **AND, OR, NOT** – conditional operators
- (condition)**?**(then)**:**(else) – if (condition) is true, select (then) else select (else). Example: the abs(x) function above can also be written as "(x> 0)?x:-x".

To check if the equation entered is valid, use the "Apply" button and the derivative will be calculated and displayed. Should the word "(undefined)" appear, it is because either a typographical error has been made, e.g. a closing bracket was left out, or an unknown function was used. In this case, you cannot use the inserted equation and close the window with "Ok", but must correct the input or click "Cancel". The "Cancel" button forfeits all changes made within the window and closes it, restoring the contents it held before it was opened. Note: The windows "input equation(s)" (Fig. C.2),

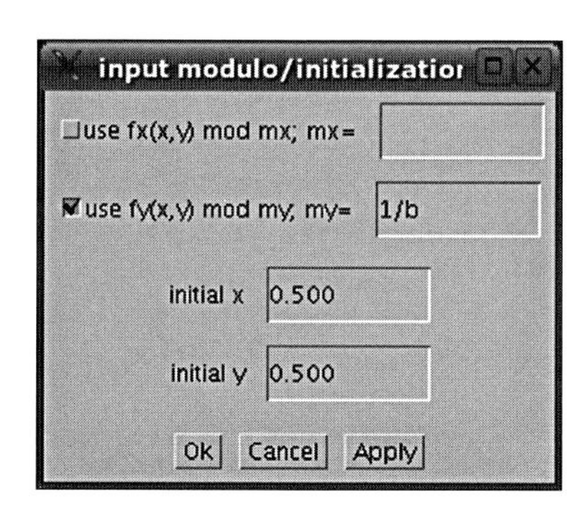

Fig. C.3 Window for "input modulo/initialization".

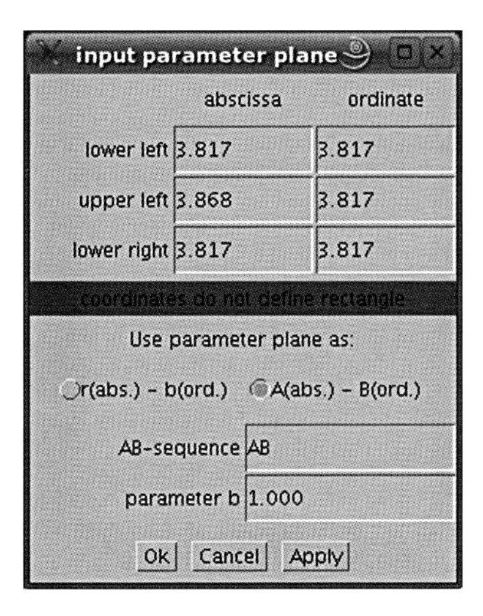

Fig. C.4 Window for "input parameter plane".

"input modulo/initializations" (Fig. C.3) or "input parameter plane (Fig. C.4) cannot be opened simultaneously. Each must be closed for another to be opened. Furthermore, these windows must be closed before the main control window can be used.

Eventual modulo operations (see e.g. Eqs. 9.21, 9.25, 9.34 and 9.48), as well as the initial values x_0 (and for 2D, also y_0), are set by clicking "input modulo/initialization", which opens the window shown in Fig. C.3. To enter a modulo value, one must first click the corresponding check-box at the left. If one clicks the check-box again, the input field is cleared. An empty field means the modulo operation is not being used. Aside from numbers, the program accepts any equation with the expressions above, excluding the variables **x** and **y**. For example, to calculate Fig. 9.14 using Eq. (9.21) one must enter "1/b" after "my=", or for Eq. (9.25) "2*pi" after "mx=". To make sure a valid value has been used, press the "Apply" button. To enter the input and close the window, click "Ok" and to restore the previous selections after closing the window, click "Cancel".

The button "input parameter plane" (Fig. C.1) opens the window shown in Fig. C.4. (For the meaning of "parameter plane", consult Section 6.3). The coordinates in the parameter plane used for the lower left, upper left and lower right corner of the λ-diagram are chosen first. Note: to define a rectangle on the plane only five coordinates need to be entered; the sixth one is set automatically, so that right angles are obtained. If all six coordinates are set, they will be corrected accordingly. If the given values fail to define a rectangle, a warning ("coordinates do not define rectangle") will be displayed. You can check the validity of the input, by clicking "Apply". Next, you must choose whether the coordinate in the parameter plane will be used for "r" as abscissa and "b" as ordinate, or if "r" will be alternated as detailed in the "AB-sequence" entered below, while the entered "b" is kept constant. Example: entering the AB-sequence AAB means that r will change periodically as AAB AAB AAB ... (see Section 6.3). As before, close the window and accept the changes by pressing the "Ok" button or close and restore the previous values by clicking "Cancel".

In the main control window (Fig. C.1), you can change the diagram width and height in pixels. Then enter the number of iterations that will be calculated after clicking the "start" button. For diagrams with AB-sequence, this is the number of times the AB-sequence (e.g. AAB or AABAB) is repeated. After pressing the "start" button, the progress is displayed and calculations can be interrupted be pressing this button again. Pressing the button a third time resumes the calculations. While calculations are interrupted, changing any entry will restart the calculations from the beginning when "start" is pressed the next time. After completing the number of iterations indicated in the main control window, the λ-diagram (e.g. Fig. C.6) will be updated. You now have the following options: (a) to calculate the next number of iterations (set in the window) by pressing the "start" button again; (b) to press "reset exponents", which sets any iterations before as "pre-iterations" and resets λ to zero; or (c) to press "reset iterations", which restarts the calculations of the iterations and λ from scratch. Resetting the number of iterations in the window is the only action that will not reset the iterations as well, while any other change of equation, parameter or else destroys the results achieved so far. The number of calculated iterations and pre-iterations is displayed at the bottom of the main control window (Fig. C.1).

C.3 Setting grey levels or colours

The colour input window (see Fig. C.5) is always visible. One can select here colours for three λ: i) the lowest (negative) λ to be considered for colouring (left of Fig. C.5); ii) an intermediate λ (it can be positive or negative; one colour is set for decreasing λ and one for increasing λ, starting from this intermediate value; center of Fig. C.5); and iii) the highest (positive) λ (right of Fig. C.5). Each colour has three sliders (upper part of Fig. C.5) for the red (top sliders), green (middle sliders) and blue (bottom sliders) colour, corresponding to the RGB-code. (Note: Although the diagrams in this book were coloured by setting the intermediate λ to zero, interesting structures are revealed by shifting this value around zero.) For a given λ, a linear interpolation between the two colours given as limits (lowest, intermediate or highest) is used if that λ is between those limits; otherwise, the colour of the nearest limit is used. The values of λ at the limits are entered below the sliders. A click on the "+" button adds 10% to the value, a click on the "-" button subtracts 10% from the value to its left.

λ may diverge to $+\infty$ or $-\infty$. These values are outside the colouring ranges, thus each merits its own colour, which is also set in the colour input window. Finally, the "democratic" shading described in Section 6.2 can be chosen. The range of calculated λ (excluding $\lambda \to \pm\infty$) is displayed at the bottom of the colour input window. After changes are made, click on "redraw" to repaint the diagram (Fig. C.6).

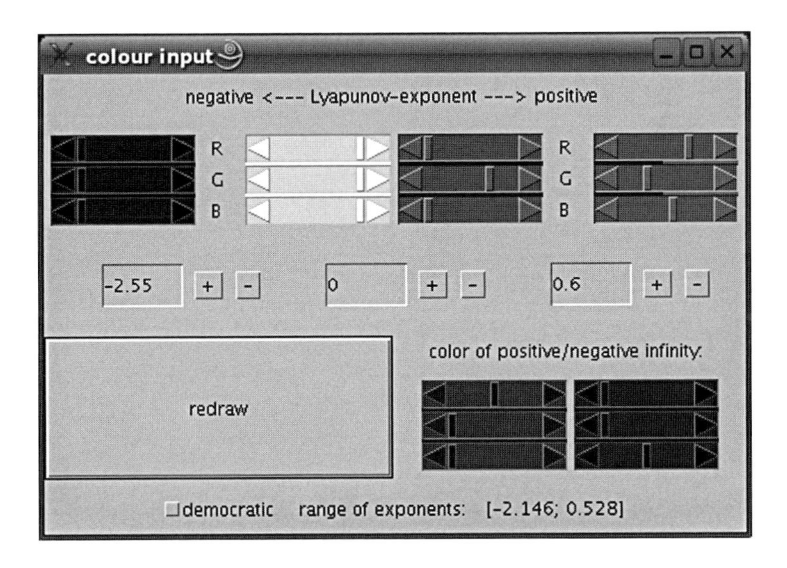

Fig. C.5 Colour input window

C.4 Blow-ups, rescalings and rotations

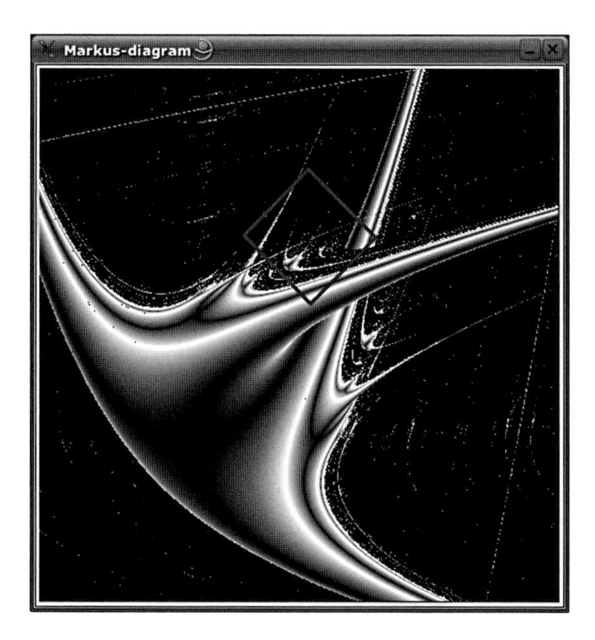

Fig. C.6 Markus-diagram window. The red rectangle is shown here as a tilted grey square in the upper middle section.

Figure C.6 shows an example of a picture in the Markus-diagram window, which is always open. With a red rectangle (initially framing the λ-diagram at its inner border), one can set coordinates in the parameter plane (zooming in on interesting details, stretching, compressing or rotating the diagram). Simply "drag" one of the four corners (not between the corners!) of this rectangle (left-click with the mouse and hold the left button pressed). The upper right corner increases or decreases the area of the rectangle; the lower left corner translates it; the other two corners rotate and change the lengths (rescale) of the respective axes. First, decrease the size of the rectangle with the upper right corner; then move the rectangle to the desired position with the lower left corner; after that, rotation and setting of axes-length with the remaining two corners is possible. Based on the dimension of the red rectangle (in particular the ratio of the two axes), one might want to change the width or height (in pixels) of the diagram in the main control window; appropriate values for these are suggested in brackets next to the input fields in that window. (Watch how they change as you manipulate the red rectangle!)

C.5 How to save, load, or exit, and how to create a personal library

The menu of the main control window allows one to:

- save the values and equations, entered within all fields except the colouring window via "File → Save Data File";
- load parameter values and equations from a previous, saved session via "File → Open Data File";
- save a λ-diagram window as an image file in the PNG format via "File → Save Image File";
- exit the program via "File → Quit Program";
- display the LIB (see Section C.6 below) via "Equations → Edit LIB", edit the equations or add new ones.

C.6 Fivefold speed with LIB

If you have sufficient programming skills and would like to perform the calculations about five times faster, you can insert the equations along with their derivatives by editing a separate file and recompiling the program to construct a library of equations, which is called LIB here. (Note: No automatic calculation of derivatives is possible when using LIB. In addition, LIB requires some entries in Java and the capability to compile a Java program.) Sample LIB files are provided on the CD.

To use the LIB check "use LIB equation No.:" in the main control window. There, choose an equation from the LIB by entering its number. The number in parentheses, which is 3 in the example in Fig. C.1 indicates the total number of equations in the LIB. The name of the currently selected equation is displayed. However, the fivefold acceleration comes at a price. The following paragraphs give instructions for the installation and use of this upgrading, which requires more special skills than before.

To be able to use the LIB, one must copy the folder "Markus_Diagrams" from the CD to a hard drive folder of one's choice. Starting the program from this directory, one can then save the changes made to the LIB. To be

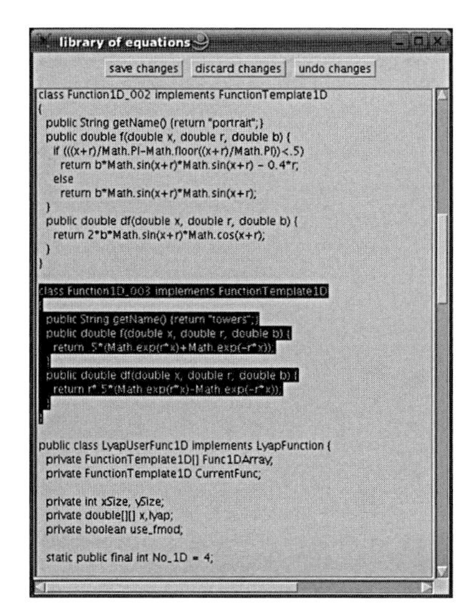

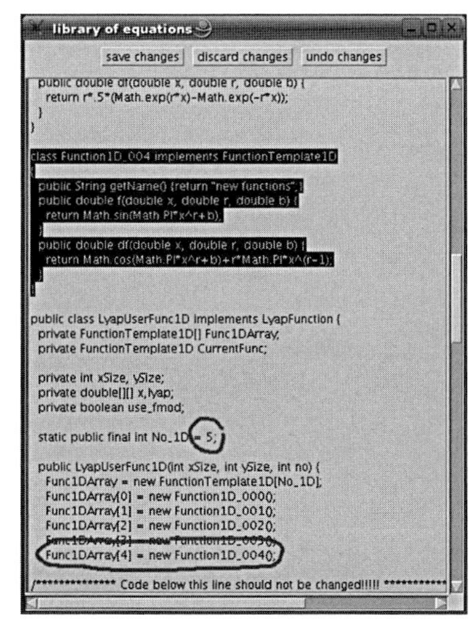

Fig. C.7 Example for editing the LIB. Left: before. Right: after. Text marked by black bars on the left is copied, inserted and edited, e.g. as marked by black bars on the right.

able to update the LIB in the program, one must have the Java Development Kit version 1.4.2 or later installed. However, an earlier version might also work. Installation files can be obtained from the Sun Microsystems homepage [129] and are also supplied on the CD.

As an example, consider the file opened when "1D" is checked and then "Equations → Edit LIB ..." is selected. The procedure for providing the program with an additional equation is illustrated in Fig. C.7. First, one must copy e.g. the part starting with "class Function1D_003 implements FunctionTemplate1D" including all 10 lines of code between the subsequent curly brackets (marked by black bars at the left of Fig. C.7). Now, in the copy, change the identifier "Function1D_003" to a new name, preferably "Function1D_004" if this is going to be the fourth 1D equation in the LIB. Next, replace the name "towers" with a new one, describing the equation to be entered. The line "public double f(...)" defines the iterative map and the chosen equation must be placed after the word "return". (The changes just described are marked by black bars at the right of Fig. C.7). For available functions in the Java programming language one can consult the Java documentation webpage [131]. Important: in Java, mathematical functions start with "Math.", e.g. sin(x) must be entered as "Math.sin(x)". After the "return" following "public double df(...)", the derivative of f(x) must be replaced as well. No automatic calculation of the derivative is performed here. The option "input equation" in the main control window may be used to calculate the correct derivative of a given equation. Next, the line "static public final int No_1D = 4" must be found, and this number "4" must be incremented by 1 (see the circled text on the right side of Fig. C.7). To conclude, after the line "Func1DArray[3] = new Function1D_003();", a new line "Func1DArray[4] = new Function1D_004();" must be appended (also circled on the right of Fig. C.7). The changes are saved and the window is closed by clicking the "save changes" button. To discard the changes and close the window, use "discard changes". Clicking the button "undo changes" reloads the file last saved.

To add equations to the 2D LIB, one must check the box "2D" in the main control window and then click "Equations → Edit LIB...". The process is analogous to the case described above, but one has to supply 6 functions (2 iterative maps called "fx()" and "fy()" and 2×2 partial derivatives called "dfxx(), dfxy(), dfyx() and dfyy()"). You should also look at the sample 2D LIB, which is available on the CD, to learn some "tricks of the trade", e.g. uses of additional parameters and auxiliary functions.

After editing the files as described above, you must leave the program and, recompile it by typing "make" in the command console. Then restart the program by entering "java -jar Lyap.jar" in the command console. The updated LIB should now be available.

For Linux/Unix systems a "Makefile" is supplied, as is a file "make.sh" for MacOS and a file "make.bat" for Windows. The latter assumes that the Java Development Kit was installed correctly. One can consult the file "README.TXT" on the CD-ROM for further hints if compilation doesn't work. Not utilizing the LIB means additional calculations and an image of equal size with identical equations takes about 5 times as long to finish. For larger images or complicated functions (e.g. Eqs. (8.17,8.18)), the use of the LIB may be unavoidable.

C.7 Larger images

By default, Java reserves 64 MB of heap memory. For images larger than 1000x1000 pixels, this is not sufficient. If one wants to calculate larger images, one can either "tile" them by calculating small images and combining them with an image editing program, or one can start the program using "java -Xmx256m -jar Lyap.jar". This reserves 256MB of heap memory to be able to calculate images up to 2000x2000 pixels. If the computer has sufficient RAM, one can use even larger numbers than 256. It is advisable not to specify more than half the available RAM, though.

Bibliography

[1] *Dynamic Pattern Formation in Chemistry and Mathematics.* A collection of scientific pictures from M. Markus, S. C. Müller, Th. Plesser and B. Hess (Exhibition Catalogue; Convex Computer Corporation, Texas and Boehringer-Ingelheim Fonds, Stuttgart, 1988)

[2] *Formbildende Dynamik in Chemie und Mathematik.* Wissenschafliche Bilder von M. Markus, S. C. Müller, Th. Plesser und B. Hess (Ausstellungskatalog; Commerzbank Dortmund, 1988)

[3] N. Gray: Critique and a science for the sake of art — Fractals and the visual-art. *Leonardo* **24** (1991) 317

[4] H.-O. Peitgen and P.H. Richter: *The Beauty of Fractals* (Springer, Berlin, 1986)

[5] E. Haeckel: *Art Forms in Nature.* (Dover Publ., New York, 1974)

[6] J. Kennedy: *On reading mathematical constructions as works of art.* Lecture at the Univ. of Lancaster (2003) and the Univ. of Helsinki (2004); `www.math.helsinki.fi/logic/people/juliette.kennedy/aest2.pdf`

[7] R. Wright: Some issues in the development of computer art as a mathematical art form. *Leonardo, Electronic Art Suppl.Issue* **1** (1988) 111

[8] M. Markus: Chaos in maps with continuous and discontinuous maxima. *Computers in Physics*, **Sept/Oct** (1990) 481

[9] M. Markus and B. Hess: Lyapunov exponents of the logistic map with periodic forcing. *Computers and Graphics* **13** (1989) 553

[10] H.-J. Hoffmann: *Verknüpfungen* (Birkhäuser-Verlag, Basel, 1992)

[11] A. M. Noll: Human or machine — A subjective comparison of Piet Mondrian's composition. *Psychological Record* **16** (1966) 1

[12] A.M. Noll: The beginnings of computer-art in the United States — A memoir. *Leonardo* **27** (1994) 39

[13] M. A. Poriau: Michael Noll's experiments with art students. *Bulletin de Psychologie* **30(14-1)** (1977) 767

[14] I.C. McManus, B. Cheema and J. Stoker: The aesthetics of composition: A study of Mondrian. *Empirical Studies of the Arts* **11** (1993) 83

[15] A. Furnham and S. Rao: Personality and the aesthetics of composition:

A study of Mondrian and Hirst. *American Journal of Psychology* **4** (2002) 233
[16] A.H. Wolach: Line spacings in Mondrian paintings and computer-generated modifications. *Journal of General Psychology* **132** (2005) 281
[17] L. Rensing: *Biologische Rhythmen und Regulation.* (G. Fischer, Stuttgart, Germany, 1973)
[18] M. Markus, D. Kuschmitz and B. Hess: Properties of strange attractors in yeast glycolysis. *Biophysical Chemistry* **22** (1985) 95
[19] M. Markus and B. Hess: Transitions between oscillatory modes in a glycolytic model system. *Proceedings of the National Academy of Sciences of the USA* **81** (1984) 4394
[20] R.B. Levien and S.M. Tan: Double pendulum — An experiment in chaos. *American Journal of Physics* **61** (1993) 1038
[21] L. A Fiedler: Cross the border, close the gap. *Playboy* (Dec. 1969), 151–258
[22] M. Markus S.C. Müller and G. Nicolis, eds.: *From Chemical to Biological Organization* (Springer, Heidelberg, Germany, 1988)
[23] A. Holden, M. Markus and H. Othmer, eds.: *Nonlinear Wave Processes in Excitable Media* (Plenum, New York, 1991)
[24] V. Pérez-Muñuzuri, V. Pérez-Villar, L.O. Chua and M. Markus, eds.: *Discretely-Coupled Dynamical Systems* (World Scientific, Singapore, 1997)
[25] M. Markus: Hallucinations: their formation in the cortex can be simulated by a computer. In: *Worlds of Consciousness* (ed. by M. Schlichting and H. Leuner, VWB-Verlag, Berlin, 1995), 131–140
[26] J.D. Murray: *Mathematical Biology* (Springer, Berlin, 1989)
[27] Jonathan Swift: *On Poetry: A Rhapsody* (I. Hoggonson, London 1733)
[28] M. Markus and B. Hess: Isotropic cellular automaton for modelling excitable media. *Nature* **347** (1990) 56
[29] M. Markus, G. Kloss and I. Kusch: Disordered waves in a homogeneous, motionless excitable medium. *Nature* **371** (1994) 402
[30] H. Poincaré: *Science et Méthode* (E. Flammarion, Paris 1908)
[31] E.N. Lorenz: Deterministic nonperiodic flow. *Journal of the Atmospheric Sciences* **20** (1963) 130
[32] C.P. Snow: *The Two Cultures* (Cambridge Univ. Press, 1993)
[33] T. Binkley: The wizard of ethereal pictures and virtual places + computer artists. SIGGRAPH '89 Art Show Catalogue, *Suppl. Issue of Leonardo* (1989), 13–20
[34] M. Markus: Are one-dimensional maps of any use in ecology? *Ecological Modelling* **63** (1992) 243
[35] R.-D. Hesch: Gesundsein und Kranksein. *Futura* **1/88** (1998) 23
[36] A. Hill: About the immediate future of modern art. *Leonardo* **20** (1987) 349
[37] T. Eagleton: *Die Illusion der Postmoderne* (Metzler, Stutgart, Germany, 1998)
[38] S. Wolfram: *Theory and Application of Cellular Automata* (World Scientific, Singapore, 1986)

[39] R. Lewin: *Complexity: Life at the Edge of Chaos* (McMillan, New York, 1992)
[40] M. M. Waldrop: *Complexity: The Emerging Science at the Edge of Order and Chaos* (Simon & Schuster, New York, 1992)
[41] M. Markus: A scientist's adventures in postmodernism (Physics, chaos, cultural convergence). *Leonardo* **33** (2000) 179
[42] Mythos aus dem Computer. *Der Spiegel* **39** (1993), 156–164; Der Kult um das Chaos. *Der Spiegel* **40** (1993), 232–241; Der Kult um das Chaos. *Der Spiegel* **41** (1993), 240–252
[43] P.H. Richter, H. Dullin and H.-O. Peitgen: Der Spiegel, das Chaos - und die Wahrheit. *Physikalische Blätter* **50** (1994) 355
[44] W.S. Franklin: Discussing the sensitivity of the atmosphere to small perturbations. *Physical Review* **6** (1898) 170
[45] H.J. Scholz: Phyllotactic iterations. *Berichte der Bunsen-Gesellschaft — Physical Chemistry, Chemical Physics* **89** (1985) 699
[46] S. Swierczkowski: On succesive settings of an arc on the circumference of a circle. *Fundamenta Mathematica* **46** (1958) 187
[47] C. Marzec and J. Kappraff: Properties of maximal spacing on a circle related to phyllotaxis and to the golden mean. *Journal of theoretical Biology* **103** (1983) 201
[48] P.H. Richter and R. Schranner: Leaf arrangement - geometry, morphogenesis and classification. *Naturwissenschaften* **65** (1978) 319
[49] H. Hayashi, S. Ishizuka, M. Ohta and K. Hirakawa: Chaotic behaviour in the Onchidium giant-neuron under sinusoidal stimulation. *Physics Letters A* **88** (1982) 435
[50] L. Glass, M.R. Guevara, J. Bélair and A. Shrier: Global bifurcations of a periodically forced biological oscillator. *Physical Review A* **29** (1984) 1348
[51] K. Tomita: Chaotic response of non-linear oscillators. *Physics Reports — Review Section of Physics Letters* **86** (1982) 113
[52] K. Tomita and T. Kai: Chaotic response of a limit-cycle. *Journal of Statistical Physics* **21** (1979) 65
[53] F. Takens: Detecting strange attractors in turbulence. *Lecture Notes in Mathematics* **898** (1981) 366
[54] J.-C. Roux, R. H. Simoyi and H. L. Swinney: Observation of a strange attractor. *Physica D* **8** (1983) 257
[55] M. Hénon: 2-dimensional mapping with a strange attractor. *Communications in Mathematical Physics* **50** (1976) 69
[56] A.N. Sharkovsky and L.O. Chua: Chaos in some 1-D discontinuous maps that appear in the analysis of electrical circuits. *IEEE Transactions on Circuits and Systems* **40** (1993) 722
[57] L. Stone: Period-doubling reversals and chaos in simple ecological models. *Nature* **365** (1993) 617
[58] M. R. Guevara, L. Glass and A. Shrier: Phase locking, period-doubling bifurcations, and irregular dynamics in periodically stimulated cardiac cells. *Science* **214** (1981) 1350

[59] A. Okubo, V. Andreasen and J. Mitchell: Chaos-induced turbulent diffusion. *Physics Letters A* **105** (1984) 169
[60] A. Erramilli, R.P. Singh and P. Pruthi: An application of deterministic chaotic maps to model packet traffic. *Queueing Systems* **20** (1995) 171
[61] J.S. Turner, J.-C. Roux, W.J. McCormick and H.L. Swinney: Alternating periodic and chaotic regimes in a chemical reaction — Experiment and theory. *Physics letters A* **85** (1981) 9
[62] K. Tomita and I. Tsuda: Towards the interpretation of Hudson's experiment on the Belousov-Zhabotinsky reaction — Chaos due to delocalization. *Progress of Theoretical Physics* **64** (1980) 1138
[63] K. Matsumoto and I. Tsuda: Noise-induced order. *Journal of Statistical Physics* **31** (1983) 87
[64] A.M. Saperstein: Chaos — A model for the outbreak of war. *Nature* **309** (1984) 303
[65] R.H. Miller: Some comments on numerical methods for chaos problems. *Celestial Mechanics and Dynamical Astronomy* **64** (1996) 33
[66] S. Yousefi, Y. Maistrenko and S. Popovych: Complex dynamics in a simple model of interdependent open economies. *Discrete Dynamics in Nature and Society* **5** (2000) 161
[67] G. Feichtinger: Nonlinear threshold dynamics: further examples for chaos in social sciences. In: *Economic Evolution and Demographic Change* (ed. by G. Haag, U. Mueller and K.G. Troitzsch, Springer, Berlin, 1992), 141–154
[68] J.A. Hołyst, M. Zebrowska and K. Urbanowicz: Observation of deterministic chaos in financial time series by recurrence plots. Can one control chaotic economy? *European Physics Journal B* **20** (2001) 531
[69] A.A. Berryman: Can economic forces cause ecological chaos? — The case of the northern California dungeness crab fishery. *Oikos* **62** (1991) 106
[70] K. Ikeda: Multiple-valued stationary state and its instability of the transmitted light by a ring cavity system. *Optics Communications* **30** (1979) 257
[71] H.D.I. Abarbanel, R. Brown, J.J. Sidorowich and L. Sh. Tsimring: The analysis of observed chaotic data in physical systems. *Reviews of Modern Physics* **65** (1993) 1331
[72] W. Lu and W. Tan: The bifurcation and chaos of laser oscillation output in a ring cavity. *Optics Communications* **61** (1987) 271
[73] L. Angelini: Antiferromagnetic effects in chaotic map lattices with a conservation law. *Physics Letters A* **307** (2003) 41
[74] J. Kockelkoren and H. Chaté: Comment on "Phase ordering in chaotic map lattices with conserved dynamics". *Physical Review E* **62** (2000) 3004
[75] C. D. Jeffries: Chaotic dynamics of instabilities in solids. *Physica Scripta* **T9** (1985) 11
[76] H.-J. Zhang, J.-H. Dai, P.-Y. Wang, C.D. Jin and B.-L. Hao: Analytical study of a bimodal map related to optical bistability. *Institute of Theoretical Physics, Academia Sinica*, AS-ITP-85-042
[77] H.-J. Zhang, P.-Y. Wang, J.-H. Dai, C.D. Jin and B.-L. Hao: Analytical

study of a bimodal mapping related to a hybrid optical bistable device using liquid crystal. *Chinese Phyics Letters* **2** (1985) 5
[78] H.-J. Zhang, J.-H. Dai, P.-Y. Wang, C.D. Jin and B.-L. Hao: Analytical study of a bimodal map related to optical bistability. *Communications in Theoretical Physics* **8** (1987) 281
[79] F.-G. Xie and B.-L. Hao: Chaos: Symbolic dynamics of the sine-square map. *Chaos, Solitons and Fractals* **3** (1993) 47
[80] J.R. Beddington, C.A. Free and J.H. Lawton: Dynamic complexity in predator-prey models framed in difference equations. *Nature* **255** (1975) 58
[81] R.V. Solé, J. Valls and J. Bascompte: Spiral waves, chaos and multiple attractors in lattice models of interacting populations. *Physics letters A* **166** (1992) 123
[82] M.P. Hassell and R.M. May: Stability of insect host-parasite models. *Journal of Animal Ecology* **42** (1973) 693
[83] K. Satoh and T. Aihara: Numerical study on a coupled logistic map as a simple model for a predator-prey system. *Journal of the Physical Society of Japan* **59** (1990) 1184
[84] G.M. Zaslavsky: Simplest case of a strange attractor. *Physics Letters A* **69** (1978) 145
[85] P. Grassberger and I. Procaccia: Measuring the strangeness of strange attractors. *Physica D* **9** (1983) 189
[86] F.C. Moon: *Chaotic Vibrations* (J. Wiley, New York, 1987)
[87] F.C. Moon: Fractal boundary for chaos in a 2-state mechanical oscillator. *Physical Review Letters* **53** (1984) 962
[88] P.J. Holmes: A nonlinear oscillator with a strange attractor. *Philosophical Transactions of the Royal Society of London A* **292** (1979) 419
[89] I Gumowski and C. Mira: Point sequences generated by two-dimensional recurrences. *Proceedings of the IFIP Congress 74*, (Stockholm, 1974), 851–855
[90] P. Cvitanović: *Universality in Chaos.* (Adam Hilger, Bristol, 1984)
[91] J. Rössler, M. Kiwi, B. Hess and M. Markus: Modulated nonlinear processes and a novel mechanism to induce chaos. *Physical Review A* **39** (1989) 5954
[92] J. Rössler, M. Kiwi and M. Markus: Ecosystems under varying ambient conditions. In: *From Chemical to Biological Organisation* (ed. by M. Markus, S.C. Müller and G. Nicolis, Springer-Verlag, Heidelberg, 1988), 319–330
[93] M. Markus, B. Hess, J. Rössler and M. Kiwi: Populations under periodically and randomly varying growth conditions. In: *Chaos in Biological Systems* (ed. by H. Degn, A.V. Holden and L.V. Olsen, Plenum. New York, 1987), 267–277
[94] M.C. de Sousa Vieira, E. Lazo and C. Tsallis: New road to chaos. *Physical Review A* **35** (1987) 945
[95] B.L. Tan and T.T. Chia: Properties of a logistic map with a sectional discontinuity. *Physical Review E* **47** (1993) 3087

[96] P. Szépfalusy and T. Tél: Properties of maps related to flows around a saddle-point. *Physica D* **16** (1985) 252
[97] R.M. May: Simple mathematical models with very complicated dynamics. *Nature* **261** (1976) 459
[98] M.C. de Sousa Vieira and C. Tsallis: The gap road to chaos and its main characteristics. In: *Instabilities and Nonequilibrium Structures II* (ed. by E. Tirapegui and D. Villarroel, Kluwer Acad. Publ., Dordrecht, 1989), 75–88
[99] L. Glass and W.-Z. Zeng: Complex bifurcations and chaos in simple theoretical models of cardiac oscillations. *Annals of the New York Academy of Sciences* **591** (1990) 316
[100] L. Glass and M.C. Mackey: *From Clocks to Chaos: The Rhythms of Life* (Princeton Univ. Press, Princeton, N.J., 1988)
[101] M.R. Guevara and L. Glass: Phase locking, period doubling bifurcations and chaos in a mathematical model of a periodically driven oscillator — A theory for the entrainment of biological oscillators and the generation of cardiac disrhythmias. *Journal of Mathematical Biology* **14** (1982) 1
[102] F. C. Hoppensteadt and J. Keener: Phase locking in biological clocks. *Journal of Mathematical Biology* **15** (1982) 339
[103] J. Keener and L. Glass: Global bifurcations of a periodically forced nonlinear oscillator. *Journal of Mathematical Biology* **21** (1984) 175
[104] E.-J. Ding: Analytic treatment of a driven oscillator with a limit-cycle. *Physical Review A* **35** (1987) 2669
[105] M.A. Aziz-Alaoui, C. Robert abd C. Grebogi: Dynamics of a Hénon-Lozi-type map. *Chaos, Solitons and Fractals* **12** (2001) 2323
[106] R. Lozi: Un attraction étrange du type attracteur de Heńon. *Journal de Physique (Paris)* **39**: Coll.C5 (1978) 9
[107] H. Degn: Strange attractors in linear period transfer functions with periodic perturbations. In: *Chemical Applications of Topology and Graph Theory* (ed. by R.B. King, Elsevier, Amsterdam, 1983), 364–370
[108] L. F. Olsen and H. Degn: Chaos in biological systems. *Quarterly Review of Biophysics* **18** (1985) 165
[109] I. Procaccia: Universal properties of dynamically complex systems — The organization of chaos. *Nature* **333** (1988) 618
[110] U. Feudel, C. Grebogi, B.R. Hunt and J.A. Yorke: Map with more than 100 coexisting low-period attractors. *Physical Review E* **54** (1996) 71
[111] C. Grebogi, S.W. McDonald, E. Ott and J.A. Yorke: Final-state sensitivity — an obstruction to predictability. *Physics Letters A* **99** (1983) 415
[112] J.C. Alexander, J.A. Yorke, Z. You and I. Kan: Riddled basins. *International Journal of Bifurcation and Chaos* **2** (1992) 795
[113] J.C. Sommerer and E. Ott: A physical system with qualitatively uncertain dynamics. *Nature* **365** (1993) 138
[114] M. Woltering and M. Markus: Riddled basins of coupled elastic arches. *Physics Letters A* **260** (1999) 453
[115] M. Woltering and M. Markus: Riddled basins in a model for the Belousov-Zhabotinsky reaction. *Chemical Physics Letters* **S 321** (2000) 473

[116] M. Woltering and M. Markus: Riddled-like basins of transient chaos. *Physical Review Letters* **84** (2000) 630

[117] Yu. Maistrenko, T. Kapitaniak and P. Szuminski: Locally and globally riddled basins in two coupled piecewise-linear maps. *Physical Review E* **56** (1997) 6393

[118] A.S. Lima, I.C. Moreira and A.M. Serra: Transitions between the tent map and the Bernoulli shift. *Phyics Letters A* **190** (1994) 403

[119] J. Kaplan and J.A. Yorke: Chaotic behaviour of multidimensional difference equations. *Springer Lecture Notes in Mathematics* **730** (1978) 228

[120] S.W. McDonald, C. Grebogi, E. Ott and J.A. Yorke: Fractal basin boundaries. *Physica D* **17** (1985) 125

[121] X. Zhang and K. Shen: Controlling spatiotemporal chaos via phase space compression. *Physical Review E* **63** (2001) 046212

[122] B.B. Mandelbrot: Fractal aspects of the iteration $z \longrightarrow \lambda\, z(1-z)$ for complex λ and z. In: Nonlinear Dynamics: *Annals of the New York Academy of Sciences* (ed. by H.G. Helleman) **357** (1980) 249

[123] Y.-C. Lai: Transition from strange nonchaotic to strange chaotic attractors. *Physical Review E* **53** (1996) 57

[124] A. Prasad, V. Mehra and R. Ramaswamy: Intermittency route to strange nonchaotic attractors. *Physical Review Letters* **79** (1997) 4127

[125] T. Yalçinkaya and Y.-C. Lai: Bifurcation to strange nonchaotic attractors. *Physical Review E* **56** (1997) 1623

[126] R. Eykholt and D.K. Umberger: Characterization of fat fractals in nonlinear dynamic-systems. *Physical Review Letters* **57** (1986) 2333

[127] M. Markus and J. Tamames: Fat fractals in Lyapunov space. In: *Fractal Horizons* (ed. by Clifford Pickover, St. Martin's Press, New York, 1996), 333–348

[128] M. Markus and B. Hess: Properties of modulated one-dimensional maps. In: *A Chaotic Hierarchy* (ed. by G. Baier and M. Klein, World Scientific, Singapore, 1991), 267–283

[129] The source for Java developers; `http://java.sun.com/`

[130] Java components for mathematics; `http://math.hws.edu/javamath/`

[131] Class Math; `http://java.sun.com/j2se/1.4.2/docs/api/java/lang/Math.html`

Descriptions of the Colour Plates

Plate 1 Simulated hallucination. The retonocortical map was applied to stationary (Turing) structures that appear in the visual cortex at low oxygen supply. Neural activation is shown black in the outer regions and red in the central region. Neural inhibition is shown red in the outer regions and yellow in the central region. (Section 4.3, [23]).

Plate 2 Equation 8.4. (cardiac oscillations). Figure foreground: chaos. Figure backround: periodicity. LL:(11, 11), UL:(11,190), LR:(0.5065, 0.505)

Plate 3 Equation 8.12. (Belousov-Zhabotinsky reaction). Dark backround of the figure: chaos. Lighter foreground of the figure: periodicty. LL:(0.505, 0.505), UL:(0.505, 0.5065), LR:(0.5065, 0.505)

Plate 4 Equations (9.1, 9.2). Chaotic attractor for $K = 0.05$, $b = 0.005$ and $r = -0.495$. The points get darker as time proceeds after the attractor is reached from its neighbourhood.

Plate 5 Equations (9.1, 9.2). Chaotic attractor for $K = 0.05$, $b = 0.006$, $r = -0.899$. The color changes from white to blue as time proceeds after the attractor is reached from its neighbourhood.

Plate 6 Equations (9.1, 9.2). Transients and chaotic attractor for $K = 0.05$, $b = 0.001$ and $r = 0.1$. Initial conditions: points within an equilateral triangle (height: 20% of the width of the figure) placed at the center of the figure. The upper "mountains" are (transient) deformations of this triangle, their height getting lower as time proceeds until they become part of the chaotic attractor (sea-like structure) at the bottom. During iterations, the color alternates between white and different shadings of blue.

Plate 7 Equations (9.1, 9.2). Transients for $K = 0.05$, $b = 0.001$ and $r = 0.1$. Initial conditions: 5000 points placed on the principal diagonal. The first points are shown red. As time proceeds the points turn yellow until they reach one point at the center of the figure.

Plate 8 Equations (9.1, 9.2). Transients for $K = 0.05$, $b = 0.005$ and $r = 0.1$. Initial conditions: 800×600 points equidistantly placed on the phase-plane. The first points are shown lower and red. As time proceeds, the points are shown higher and turn yellow until they reach four points on the plane.

Plate 9 Equation (9.4). r:AB AB..., $x_0 = 0.5$. Black to yellow: from λ from its minimum to 0. Black to red: λ from 0 to its maximum. LL:(2,2), UL:(2,4), LR:(4,2)

Plate 10 Equation (9.4). r:AB AB..., Colouring as in Plate 9. $x_0 = 0.5$. LL:(3.817, 3.817), UL:(3.817, 3.868), LR:(3.868, 3.817)

Plate 11 Equation (9.4). r:AAABB AAABB..., $x_0 = 0.5$. Picture foreground: $\lambda < 0$. Background: $\lambda > 0$. LL=(3.8358, 3.59), UL:(3.8358, 3.6), LR:(3.8405, 3.59)

Plate 12 Equation (9.4). r:ABAABBAAABBB ABAABBAAABBB..., $x_0 = 0.5$. Black to light blue: λ from its minimum to 0. White to dark blue: λ from 0 to its maximum. LL:(3.18, 3.73), UL:(3.18, 3.96), LR:(3.52, 3.73)

Plate 13 Equation (9.4). r: A^5B^5 A^5B^5..., $x_0 = 0.5$. Black to white: λ from its minimum to 0. Light to dark blue: λ from 0 to its maximum. LL:(3.44, 3.141), UL:(3.6435, 3.27), LR:(3.625, 3.849)

Plate 14
Equation (9.4). r: ABAABBA^3B^3AABBAB ABAABBA^3B^3AABBAB..., $x_0 = 0.8315$. Black to green: λ from its minimum to 0. Yellow to green: λ from 0 to its maximum. LL:(1.569, 3.793), UL:(1.569, 3.9385), LR:(1.602, 3.793)

Plate 15 Equation (9.4). r:AABABAB AABABAB..., $x_0 = 0.5$. Black to yellow: λ from its minimum to 0. Black to blue: λ from 0 to its maximum. LL:(2.759, 3.21), UL:(2.759, 4), LR:(3.744, 3.21)

Plate 16 Equation (9.4). r:AABAB AABAB..., $x_0 = 0.5$. Blue to white: λ from its minimum to 0. Black to blue: λ from 0 to its maximum. LL:(3.45, 3.54), UL:(3.45, 3.67), LR:(3.61, 3.54)

Plate 17 Equation (9.4). r:$B^{12}A$ $B^{12}A$..., $x_0 = 0.5$. Colouring as in Plate 13. LL:(3.51, 3.24), UL:(3.51, 3.59), LR:(4., 3.24)

Plate 18 Equation (9.4). r:A^6B^6 A^6B^6..., $x_0 = 0.5$. Blue to white λ from its minimum to 0. Dark blue: $\lambda > 0$. LL:(2.52, 3.46), UL:(2.52, 4), LR:(3.65, 3.46)

Plate 19 Equation (9.4). r: $B^7A^2B^9(BA)^9A^7B^2A^7$..., $x_0 = 0.8315$. Red to black: $\lambda > 0$. Rest (foreground): $\lambda < 0$. LL:(3.769, 1.155), UL:(3.769, 1.197), LR:(3.8175, 1.155)

Plate 20 Equation (9.6). $\alpha = 0.9935$. r:AB AB..., $x_0 = 0.7$. White to blue: λ from its minimum to 0. White to blue: λ from 0 to its maximum. LL:(3.815, 3.838), UL:(3.848, 3.871), LR:(3.839, 3.814)

Plate 21 Equation (9.6). $\alpha = 0.907$. r:BBABABA BBABABA..., $x_0 = 0.499$. White over brown to black: λ from its minimum to 0. Brown to black: λ from 0 to its maximum. LL:(3.8175, 3.8138), UL:(3.8175, 3.8181), LR:(3.8097, 3.8138)

Plate 22 Equation (9.6). Section of Plate 21. Yellow over red to black: λ from its minimum to 0. Black over red to black: λ from 0 to its maximum. LL:(3.8155, 3.8144), UL:(3.8155, 3.8177), LR:(3.8121, 3.8144)

Plate 23 Equation (9.6). $\alpha = 0.908$. r:AB AB..., $x_0 = 0.499$. Yellow over red to black: λ from its minimum to 0. Yellow over red to black: λ from 0 to its maximum. LL:(3.799, 3.8143), UL: (3.8115, 3.828), LR:(3.8169, 3.7980)

Plate 24 Equation (9.6). $\alpha = 0.907$. r:AB AB..., $x_0 = 0.499$. Changes between yellow, red and black: $\lambda < 0$. Red to black: λ from 0 to its maximum. LL:(3.8109, 3.8109), UL:(3.8109, 3.8207), LR:(3.8207, 3.8109)

Plate 25 Equation (9.47). $b = 1.95$. r: A^6B^6 A^6B^6..., $x_0 = 0$. Black to red: λ from its minimum to 0. Yellow: $\lambda > 0$. LL:(3.11, 2.02), UL:(2.21, 1.13), LR:(2.033, 3.109)

Plate 26 Equation (9.47). $b = 1.7$. r:$A^{10}B^{10}$ $A^{10}B^{10}$..., $x_0 = 0$. White to grey: λ from its minimum to 0. Black to blue: λ from 0 to its maximum. LL:(0.203, 2.748), UL:(0.167, 2.784), LR:(0.264, 2.809)

Plate 27 Equation (9.47). $b = 2.5$. r=AB AB..., $x_0 = 0$. Black to dark blue: λ from its minimum to 0. Light blue to white: λ from 0 to its maximum White to grey: λ from its minimum to 0. LL:(0,0), UL:(0,10), LR:(0,8.8)

Plate 28 Equation (9.47). $b = 1.95$. r:A^{6}B^6 A^{6}B^6..., $x_0 = 0$. Dark green to white: λ from its minimum to 0. Dark green: $\lambda > 0$. LL:(0.705, −0.453), UL:(1.15, −0.847), LR:(0.075, −1.165)

Plate 29 Equation (9.47). $b = 2$. r:A^{10}B^{10} A^{10}B^{10}..., $x_0 = 1$. White to brown: λ from its minimum to 0. Black over green to light blue: λ from 0 to its maximum. LL:(0.634, 0.127), UL:(1.06, 0.361), LR:(0.408, 0.538)

Plate 30 Equation (9.47). $b = 2.6$. r:A^{6}B^6 A^{6}B^6..., $x_0 = 0$. Orange: $\lambda < 0$. Black to blue: λ from 0 to its maximum. LL:(3.145, 1.715), UL:(2.966, 1.843), LR:(3.283, 1.909)

Plate 31 Equation (9.47). $b = 2.7$. r:A^{6}B^6 A^{6}B^6..., $x_0 = 0$. White over ocher to black: λ from its minimum to 0. Blue to black: λ from 0 to its maximum. LL:(1.225, 1.015), UL:(0.985, 0.775), LR:(1.015, 1.225)

Plate 32 Equation (9.47). $b = 2.8$. r:AB AB..., $x_0 = 0$. Black to yellow: λ from its minimum to 0. Black: $\lambda > 0$. LL:(0,0), UL:(0,10), LR:(0,8.8)

Plate 33 Equation (9.47). Yellow to orange: λ from its minimum to 0. Black: $\lambda > 0$. (See end of this list).

Plate 34 Equation (9.47). $b = 2.05$. r:A^{3}B^3 A^{3}B^3..., $x_0 = 0$. Magenta to black: λ from its minimum to 0. Dark to light blue: λ from 0 to its maximum. LL:(2.157, 3.11), UL:(1.282, 2.235), LR:(3.11, 2.158)

Note: The data for Plates 33 and 35–44 was lost along with PCs in a wretched burglary committed at the Max Planck Institute, Dortmund, in the summer of 1996. The reader may encounter these pictures by exploring the parameter space, turning the mishap into a challenge. Plates 35–44 were obtained with Equation 9.48, setting $n \to -\infty$ and $\gamma = \mu = 1$.

Plate 1

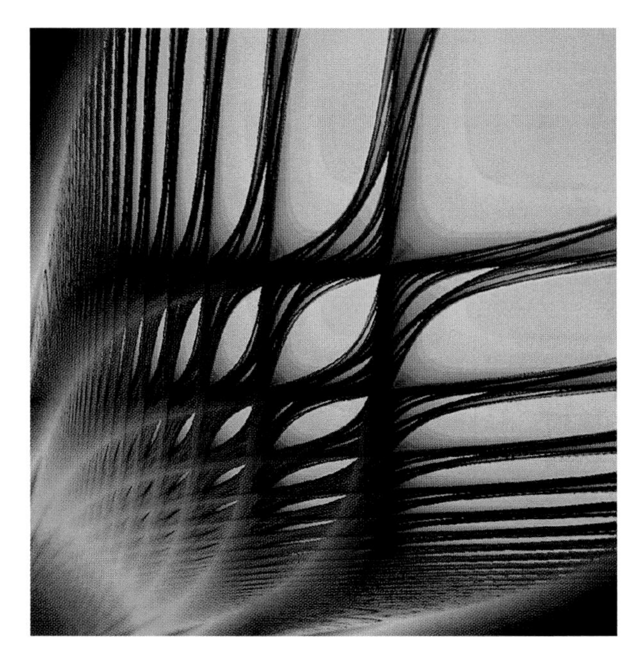

Plate 2

Plate 3

Plate 4

Plate 5

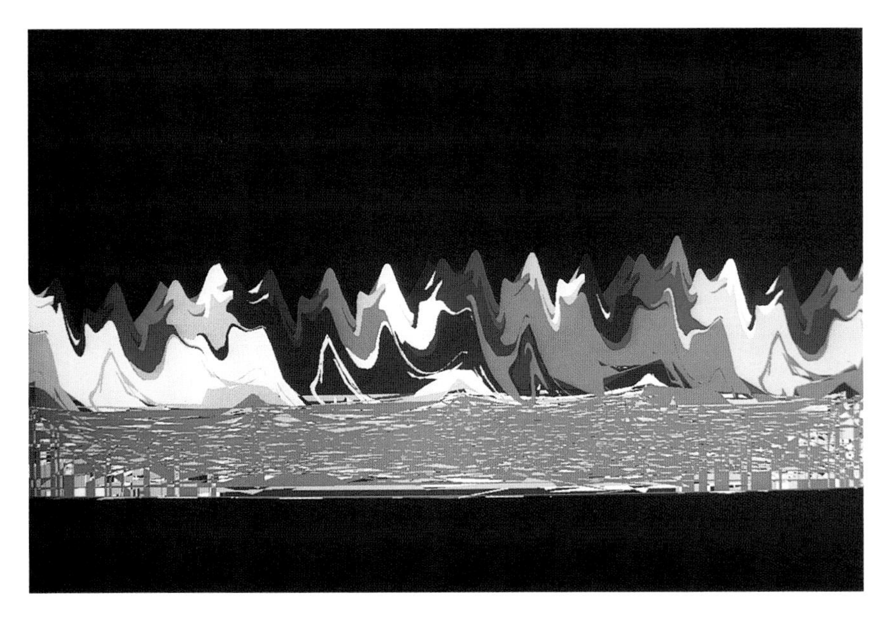

Plate 6

Plate 7

Plate 8

Plate 9

Plate 10

Plate 11

Plate 12

Plate 13

Plate 14

Plate 15

Plate 16

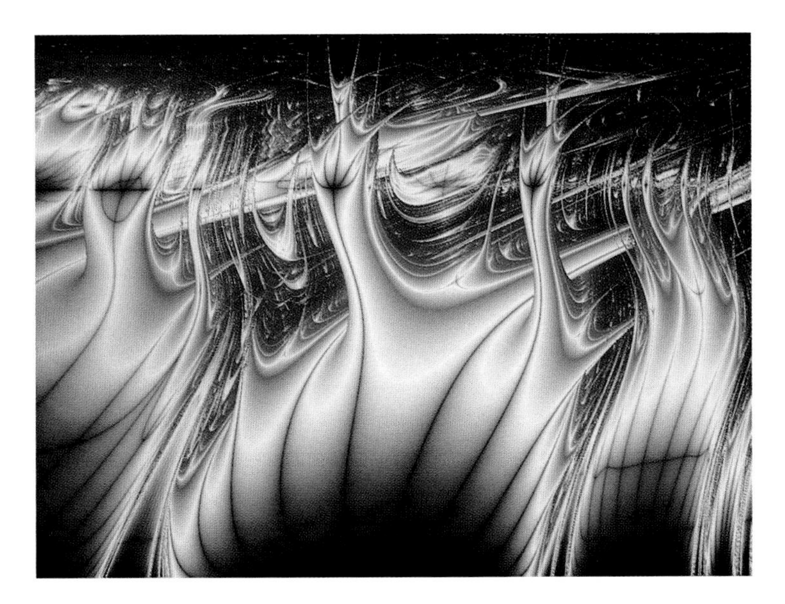

Plate 17

Plate 18

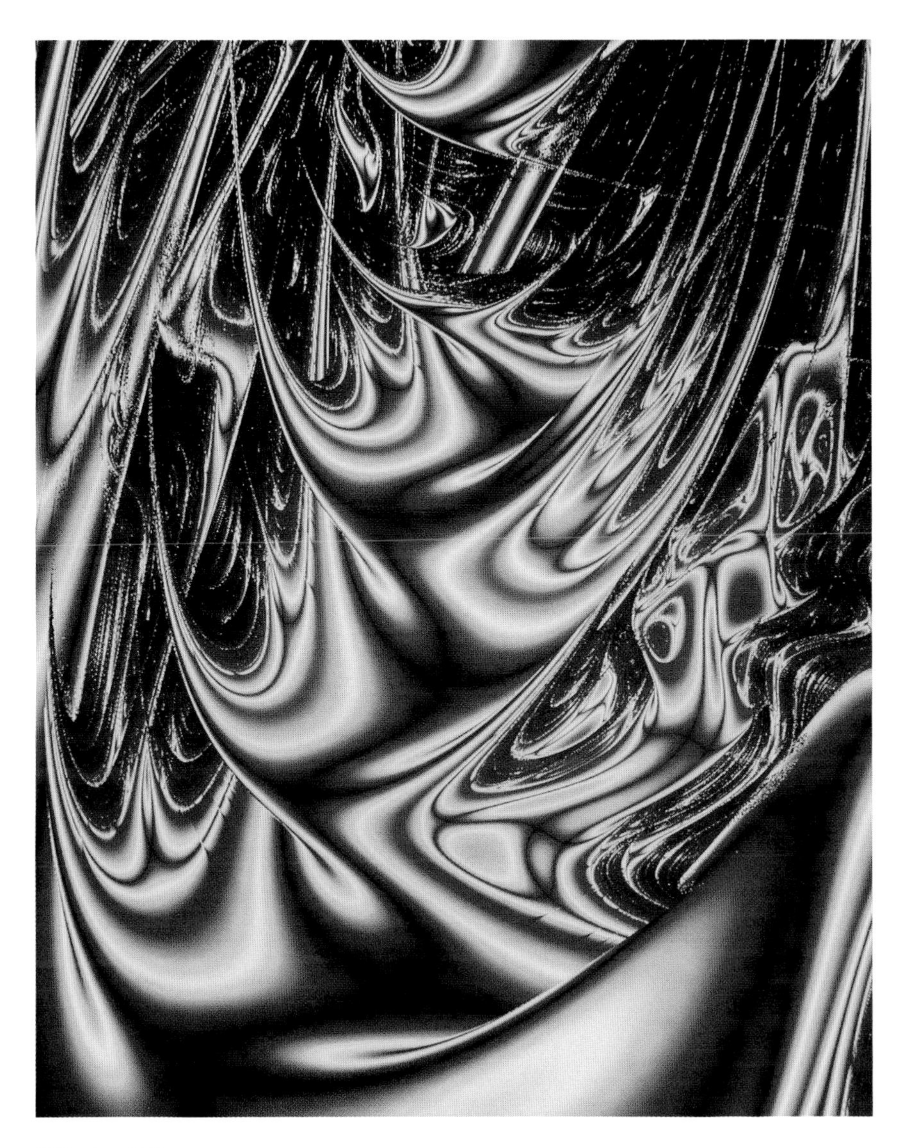

Plate 19

Plate 20

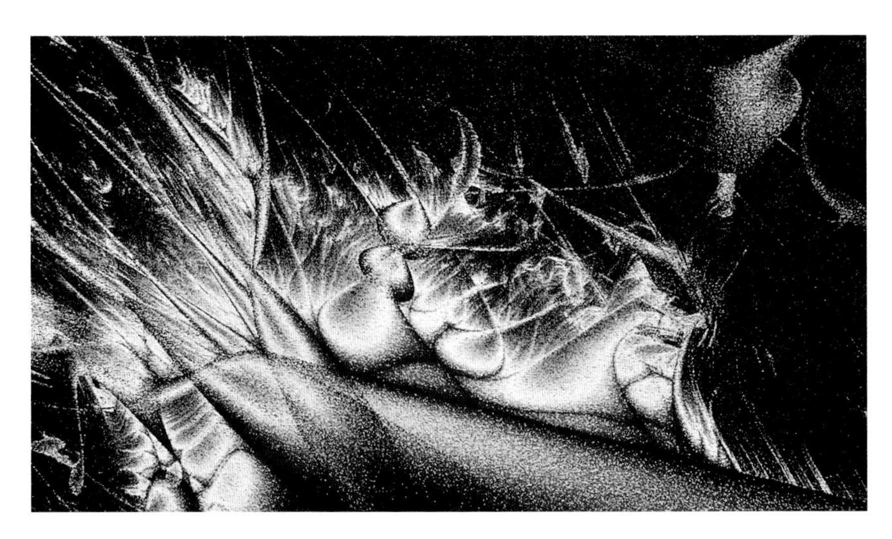

Plate 21

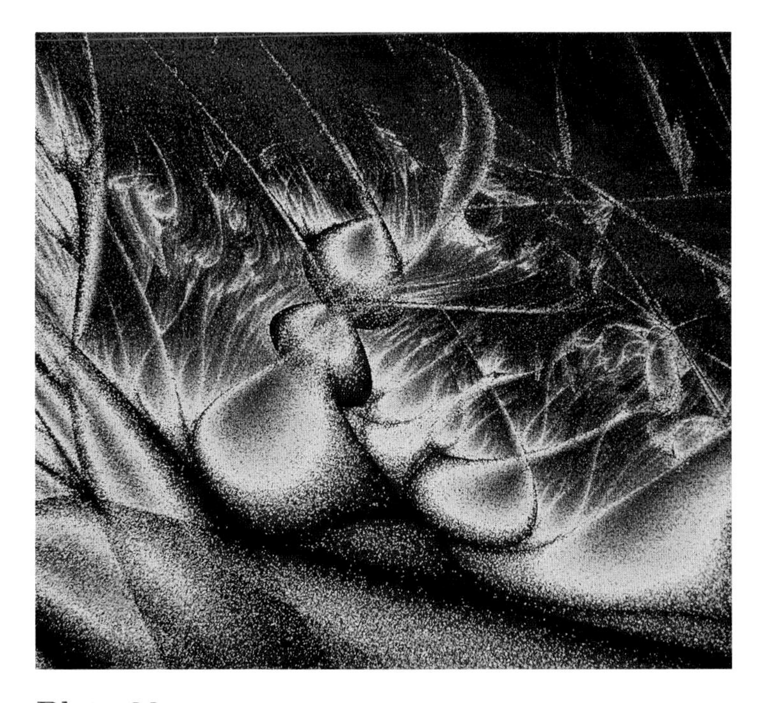

Plate 22

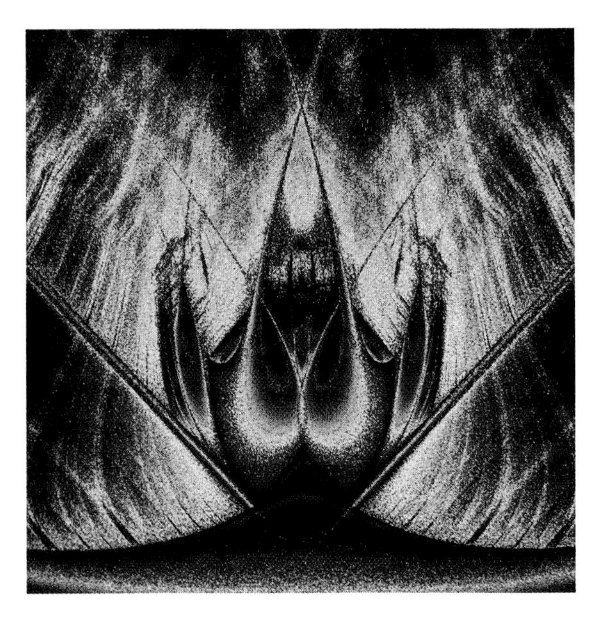

Plate 23

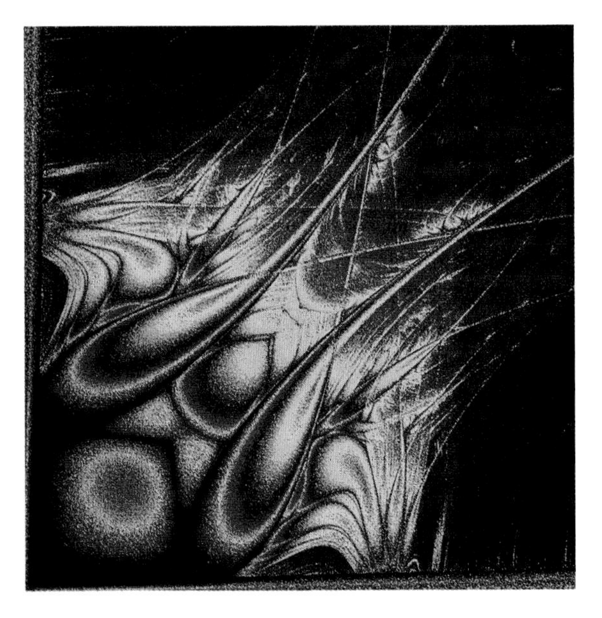

Plate 24

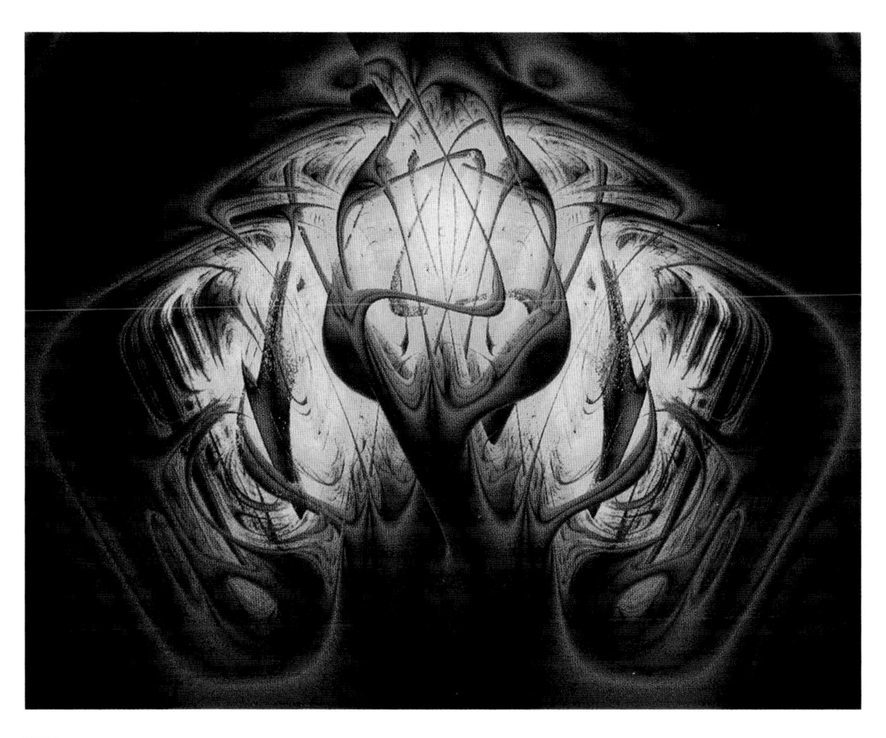

Plate 25

Plate 26

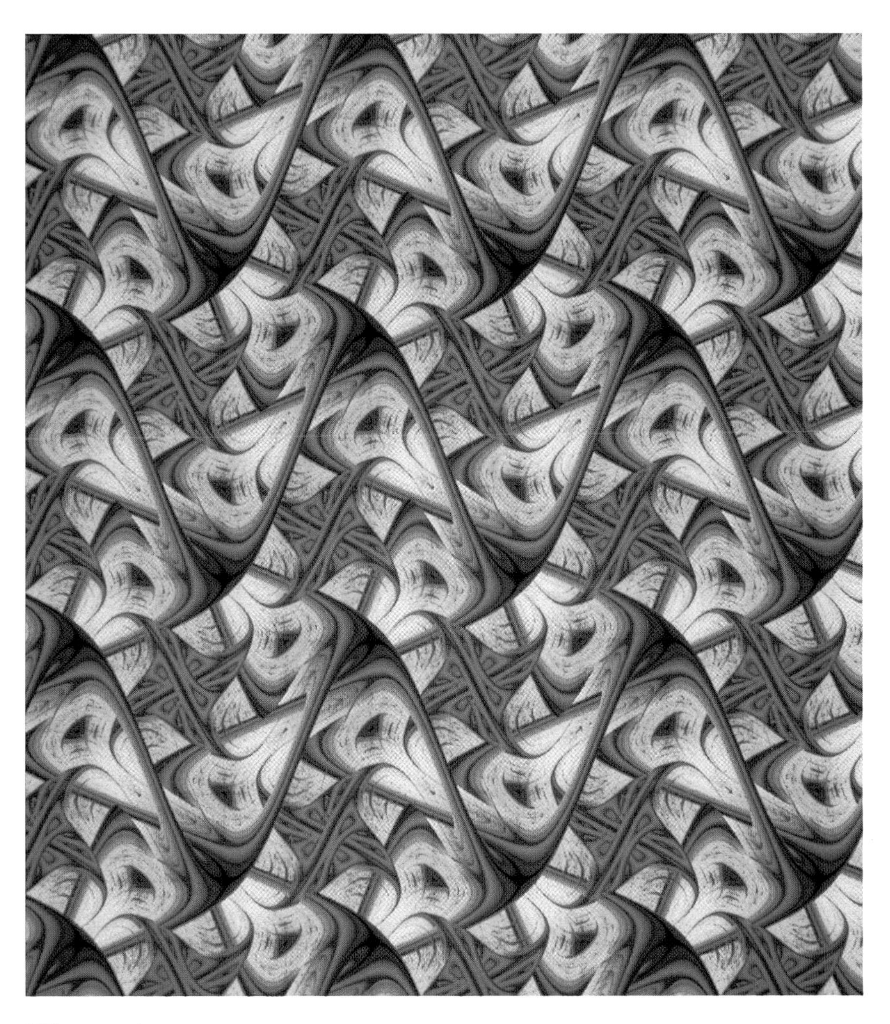

Plate 27

Plate 28

Plate 29

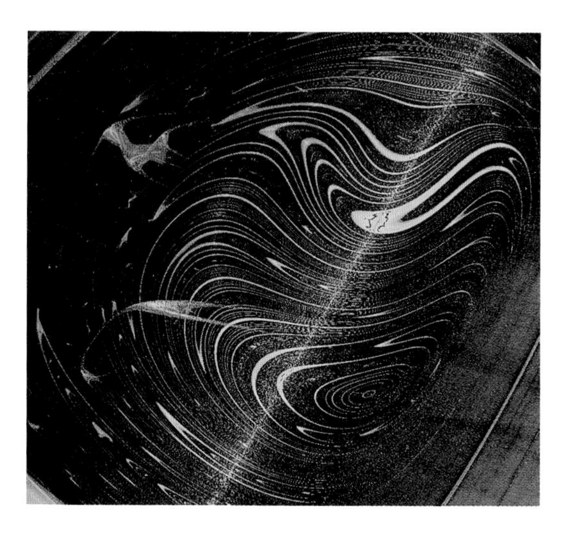

Plate 30

Plate 31

Plate 32

Plate 33

Plate 34

Plate 35

Plate 36

Plate 37

Plate 38

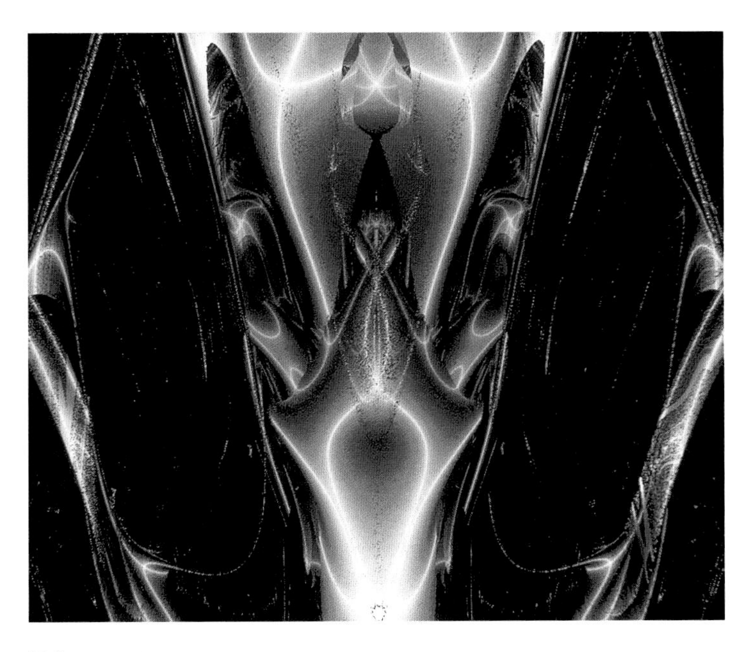

Plate 39

Plate 40

Plate 41

Plate 42

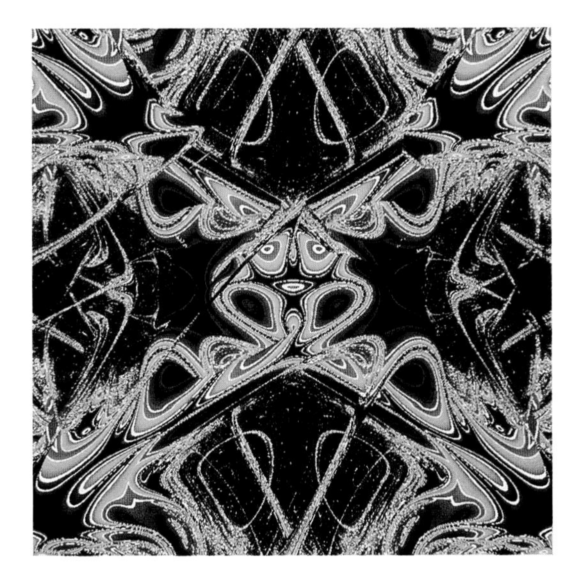

Plate 43

Plate 44

Index